The Gradient Test

The Gradient Test: Another Likelihood-Based Test

Artur J. Lemonte
Departamento de Estatística, CCEN
Universidade Federal de Pernambuco
Cidade Universitária, Recife/PE, 50740-540, Brazil

ELSEVIER

AMSTERDAM • BOSTON • HEIDELBERG • LONDON
NEW YORK • OXFORD • PARIS • SAN DIEGO
SAN FRANCISCO • SINGAPORE • SYDNEY • TOKYO
Academic Press is an imprint of Elsevier

Academic Press is an imprint of Elsevier
125 London Wall, London, EC2Y 5AS, UK
525 B Street, Suite 1800, San Diego, CA 92101-4495, USA
50 Hampshire Street, 5th Floor, Cambridge, MA 02139, USA
The Boulevard, Langford Lane, Kidlington, Oxford OX5 1GB, UK

ISBN: 978-0-12-803596-2

British Library Cataloguing in Publication Data
A catalogue record for this book is available from the British Library

Library of Congress Cataloging-in-Publication Data
A catalog record for this book is available from the Library of Congress

For information on all Academic Press publications
visit our website at http://store.elsevier.com/

To George Terrell

CONTENTS

LIST OF FIGURES

LIST OF TABLES

PREFACE

Likelihood-based methods in the statistic literature have long data and started with Ronald Aylmer Fisher, who proposed the likelihood function as a means of measuring the relative plausibility of various values of parameters by comparing their likelihood ratios. The statistical inference based directly on the likelihood function was only intensified in the period 1930–40 thanks to Sir Ronald Fisher. Nowadays, the basic ideas for likelihood are outlined in many books. Basically, the likelihood function for a parametric model is viewed as a function of the parameters in the model with the data held fixed. This function has also been extended and generalized to semi-parametric and non-parametric models, and various pseudolikelihood functions have been proposed for more complex models. In short, it is evident that the likelihood function provides the foundation for the study of theoretical statistics, and for the development of statistical methodology in a wide range of applications.

The large-sample tests usually employed for testing hypotheses in parametric models are the following three likelihood-based tests: the likelihood ratio test, the Wald test, and the score test. The score test is often known as the Lagrange multiplier test in econometrics. The large-sample tests that use the likelihood ratio, Wald, and Rao score statistics were proposed by Samuel S. Wilks in 1938, Abraham Wald in 1943, and Calyampudi R. Rao in 1948, respectively. It worth emphasizing that the likelihood ratio, Wald, and score statistics are covered in almost every book on statistical inference and provide the base for testing inference in practical applications. These three statistics for testing composite or simple null hypothesis $\mathcal{H}_0$ against an alternative hypothesis $\mathcal{H}_a$, in regular problems, have a χ_k^2 null distribution asymptotically, where k is the difference between the dimensions of the parameter spaces under the two hypotheses being tested.

After more than 50 years since the last likelihood-based large-sample test (ie, the test that uses the Rao score statistic) was proposed, a new likelihood-based large-sample test was developed, and it was introduced by George R. Terrell in 2002. The new test statistic proposed by him is quite simple to be computed when compared with the other three classic statistics and hence may be an interesting alternative to the usual large-sample test statistics for testing hypotheses in parametric models. He named it as the *gradient statistic*. An advantage of the gradient statistic over the Wald and score statistics is that it does not involve knowledge of an information matrix, neither expected nor observed. An interesting result about the gradient statistic is that it shares the same first order asymptotic properties with the likelihood ratio, Wald, and score statistics; that is, to the first order of approximation, the gradient statistic has the same asymptotic distributional properties as that of the likelihood ratio, Wald, and score statistics either under the null hypothesis or under a sequence of Pitman alternatives.

The material we present in this book is a compilation of analytical results and numerical evidence available in the literature on the gradient statistic. Only the last chapter of the book deals with new results regarding the gradient test statistic; that is, we propose in the last chapter a robust gradient-type statistic which is robust to outlying observations. Our goal is to present, in a coherent way, the main results for the gradient test statistic considered in the statistic literature so far. We also would like to point out that the details involved in many of the derivations were not included in the text since we intend to provide readers with a concise monograph. Further details can be found in the references listed at the end of the book. In short, the main aim of the current book is to divulge/disseminate the new large-sample test statistic to the users and to the researchers who devote their researches in likelihood-based theory.

Finally, the author would like to thank Francisco Cribari-Neto from Federal University of Pernambuco (Brazil) for reading the book very carefully and for providing many suggestions. The author also thanks Alexandre B. Simas from Federal University of Paraíba (Brazil) for helping to prove some results in the last chapter. The financial support from CNPq (Brazil) and FACEPE (Pernambuco, Brazil) is also gratefully acknowledged.

Artur J. Lemonte
Recife

CHAPTER 1

The Gradient Statistic

1.1 BACKGROUND

It is well-known that the likelihood ratio (LR), Wald, and Rao score test statistics are the most commonly used statistics for testing hypotheses in parametric models [1–3]. These statistics are widely used in disciplines as widely varied as agriculture, demography, ecology, economics, education, engineering, environmental studies, geography, geology, history, medicine, political science, psychology, sociology, etc., to make inference about specific parametric models. To emphasize their key role in statistical inference, Rao in Ref. [4] named them "the Holy Trinity." These three statistics are covered in almost every book on statistical inference and hence we shall briefly discuss about them in what follows.

Let $x_1, \ldots, x_n$ be an n-dimensional sample with each x_l ($l = 1, \ldots, n$) having probability density function $f(\cdot;\boldsymbol{\theta})$, which depends on a p-dimensional vector of unknown parameters $\boldsymbol{\theta} = (\theta_1, \ldots, \theta_p)^\top$. We assume that $\boldsymbol{\theta} \in \boldsymbol{\Theta}$, where $\boldsymbol{\Theta} \subseteq \mathbb{R}^p$ is an open subset of the Euclidean space. Let

The Gradient Test. http://dx.doi.org/10.1016/B978-0-12-803596-2.00001-6

$Lik(\boldsymbol{\theta})$ be the likelihood function (often simply the likelihood), which is given by

$$Lik(\boldsymbol{\theta}) = \prod_{l=1}^{n} f(x_l; \boldsymbol{\theta}).$$

The likelihood is viewed as a function of the unknown parameter $\boldsymbol{\theta}$ for a given data set. It is often numerically convenient to work with the natural logarithm of the likelihood function, the so-called log-likelihood function:

$$\ell(\boldsymbol{\theta}) = \log Lik(\boldsymbol{\theta}) = \sum_{l=1}^{n} \log f(x_l; \boldsymbol{\theta}).$$

The first derivative of the above function is called score function (or efficient score), and it is given by

$$\boldsymbol{U}(\boldsymbol{\theta}) = \frac{\partial \ell(\boldsymbol{\theta})}{\partial \boldsymbol{\theta}}.$$

Note that the score function is a p-vector of first partial derivatives, one for each element of $\boldsymbol{\theta}$. It is possible to show that

$$\mathbb{E}[\boldsymbol{U}(\boldsymbol{\theta})] = \mathbf{0}_p,$$

where $\mathbf{0}_k$ denotes a k-dimensional vector of zeros; that is, the score function has mean zero. The Fisher information matrix is given by

$$\boldsymbol{K}(\boldsymbol{\theta}) = \mathbb{E}[\boldsymbol{U}(\boldsymbol{\theta})\boldsymbol{U}(\boldsymbol{\theta})^\top].$$

Under mild regularity conditions, it can be shown that $\boldsymbol{K}(\boldsymbol{\theta}) = -\mathbb{E}[\partial^2 \ell(\boldsymbol{\theta})/\partial\boldsymbol{\theta}\partial\boldsymbol{\theta}^\top]$. Notice that the Fisher information matrix equals the score function variance-covariance.

Consider the problem of testing the simple null hypothesis $\mathcal{H}_0 : \boldsymbol{\theta} = \boldsymbol{\theta}_0$ against the two-sided alternative hypothesis $\mathcal{H}_a : \boldsymbol{\theta} \neq \boldsymbol{\theta}_0$, where $\boldsymbol{\theta}_0$ is a fixed p-dimensional vector. The LR (S_{LR}), Wald (S_{W}), and score (S_{R}) statistics for testing $\mathcal{H}_0 : \boldsymbol{\theta} = \boldsymbol{\theta}_0$ are given by

$$\begin{aligned} S_{\text{LR}} &= 2[\ell(\hat{\boldsymbol{\theta}}) - \ell(\boldsymbol{\theta}_0)], \\ S_{\text{W}} &= (\hat{\boldsymbol{\theta}} - \boldsymbol{\theta}_0)^\top \boldsymbol{K}(\hat{\boldsymbol{\theta}})(\hat{\boldsymbol{\theta}} - \boldsymbol{\theta}_0), \\ S_{\text{R}} &= \boldsymbol{U}(\boldsymbol{\theta}_0)^\top \boldsymbol{K}(\boldsymbol{\theta}_0)^{-1} \boldsymbol{U}(\boldsymbol{\theta}_0), \end{aligned}$$

where $\hat{\boldsymbol{\theta}} = (\hat{\theta}_1, \ldots, \hat{\theta}_p)^\top$ is the maximum likelihood estimator (MLE) of $\boldsymbol{\theta} = (\theta_1, \ldots, \theta_p)^\top$, which can be obtained as the solution of the equation $\boldsymbol{U}(\hat{\boldsymbol{\theta}}) = \mathbf{0}_p$. Under certain regularity conditions and under the null hypothesis, the test statistics S_{LR}, S_{W}, and S_{R} have a central χ^2 distribution with p degrees of freedom (χ_p^2) up to an error of order $O(n^{-1})$; see, for example, Sen and Singer [5]. The null hypothesis is rejected for a given nominal level, γ say, if the observed value of the test statistic exceeds the upper $100(1-\gamma)\%$ quantile of the χ_p^2 distribution.

An alternative version of the Wald statistic is

$$S_{\text{W}}^{\dagger} = (\hat{\boldsymbol{\theta}} - \boldsymbol{\theta}_0)^{\top} \boldsymbol{K}(\boldsymbol{\theta}_0)(\hat{\boldsymbol{\theta}} - \boldsymbol{\theta}_0),$$

where we replace $\hat{\boldsymbol{\theta}}$ with $\boldsymbol{\theta}_0$ in $\boldsymbol{K}(\boldsymbol{\theta})$. Another alternative version for the Wald statistic with the same asymptotic distribution uses the information matrix estimated from the data, that is, it uses the observed information matrix instead of the expected information matrix. Similarly we can define alternative versions for the score statistic with the same asymptotic distribution.

1.2 THE GRADIENT TEST STATISTIC

The gradient statistic was proposed by Terrell [6]. Terrell's idea to construct a new test statistic was very simple and the new statistic arises from the score and Wald statistics. In order to make the connection among these statistics explicit, choose any square root, $\boldsymbol{L}(\boldsymbol{\theta})$ say, of the Fisher information matrix $\boldsymbol{K}(\boldsymbol{\theta})$; that is, find a square solution to $\boldsymbol{L}(\boldsymbol{\theta})^{\top}\boldsymbol{L}(\boldsymbol{\theta}) = \boldsymbol{K}(\boldsymbol{\theta})$. Note that we can express

$$\begin{aligned}
S_{\text{R}} &= \boldsymbol{U}(\boldsymbol{\theta}_0)^{\top} (\boldsymbol{L}(\boldsymbol{\theta}_0)^{\top}\boldsymbol{L}(\boldsymbol{\theta}_0))^{-1} \boldsymbol{U}(\boldsymbol{\theta}_0) \\
&= [(\boldsymbol{L}(\boldsymbol{\theta}_0)^{-1})^{\top} \boldsymbol{U}(\boldsymbol{\theta}_0)]^{\top} (\boldsymbol{L}(\boldsymbol{\theta}_0)^{-1})^{\top} \boldsymbol{U}(\boldsymbol{\theta}_0), \\
S_{\text{W}}^{\dagger} &= (\hat{\boldsymbol{\theta}} - \boldsymbol{\theta}_0)^{\top} \boldsymbol{L}(\boldsymbol{\theta}_0)^{\top} \boldsymbol{L}(\boldsymbol{\theta}_0)(\hat{\boldsymbol{\theta}} - \boldsymbol{\theta}_0) \\
&= [\boldsymbol{L}(\boldsymbol{\theta}_0)(\hat{\boldsymbol{\theta}} - \boldsymbol{\theta}_0)]^{\top} \boldsymbol{L}(\boldsymbol{\theta}_0)(\hat{\boldsymbol{\theta}} - \boldsymbol{\theta}_0).
\end{aligned}$$

From the above expressions, define $\boldsymbol{P}_1 = (\boldsymbol{L}(\boldsymbol{\theta}_0)^{-1})^{\top}\boldsymbol{U}(\boldsymbol{\theta}_0)$ and $\boldsymbol{P}_2 = \boldsymbol{L}(\boldsymbol{\theta}_0)(\hat{\boldsymbol{\theta}} - \boldsymbol{\theta}_0)$. It follows that the standardized vectors $\boldsymbol{P}_1$ and $\boldsymbol{P}_2$ each have asymptotic multivariate normal distribution: $\boldsymbol{P}_1 \overset{a}{\sim} \mathcal{N}_p(\boldsymbol{0}_p, \boldsymbol{I}_p)$ and $\boldsymbol{P}_2 \overset{a}{\sim} \mathcal{N}_p(\boldsymbol{0}_p, \boldsymbol{I}_p)$, when n is large, $\overset{a}{\sim}$ denoting approximately distributed, and $\boldsymbol{I}_k$ denotes the identity matrix of order k. Also, $\boldsymbol{P}_1$ and $\boldsymbol{P}_2$ converge to each other in probability: $\boldsymbol{P}_1 - \boldsymbol{P}_2 \overset{\mathbb{P}}{\to} \boldsymbol{0}_p$. It then follows that the inner product of these standardized vectors has asymptotically χ_p^2 distribution:

$$\begin{aligned}
\boldsymbol{P}_1^{\top}\boldsymbol{P}_2 &= [(\boldsymbol{L}(\boldsymbol{\theta}_0)^{-1})^{\top}\boldsymbol{U}(\boldsymbol{\theta}_0)]^{\top}[\boldsymbol{L}(\boldsymbol{\theta}_0)(\hat{\boldsymbol{\theta}} - \boldsymbol{\theta}_0)] \\
&= \boldsymbol{U}(\boldsymbol{\theta}_0)^{\top}(\hat{\boldsymbol{\theta}} - \boldsymbol{\theta}_0).
\end{aligned}$$

We have the following definition.

Definition 1.1. The gradient statistic, S_{T}, for testing the null hypothesis $\mathcal{H}_0 : \boldsymbol{\theta} = \boldsymbol{\theta}_0$ against the two-sided alternative hypothesis $\mathcal{H}_{\mathrm{a}} : \boldsymbol{\theta} \neq \boldsymbol{\theta}_0$ is given by

$$S_{\mathrm{T}} = \boldsymbol{U}(\boldsymbol{\theta}_0)^{\top}(\hat{\boldsymbol{\theta}} - \boldsymbol{\theta}_0). \tag{1.1}$$

Notice that S_{T}, the gradient statistic, has a very simple form and does not involve the information matrix, neither expected nor observed, unlike its progenitors (the Rao score and Wald statistics). The statistic S_{T} is simply the inner product of two vectors. Also, it is symmetric in the hypothesized $\boldsymbol{\theta}_0$ and observed $\hat{\boldsymbol{\theta}}$ (remember that $\boldsymbol{U}(\hat{\boldsymbol{\theta}}) = \mathbf{0}_p$). Unlike the LR and score statistics and like the Wald statistic, the gradient statistic is not invariant under nonlinear reparameterizations of the parameter vector $\boldsymbol{\theta}$ used to describe the model, or equivalently it depends on how the hypotheses are formulated. It means that the observed value of the gradient statistic may vary on how the null hypothesis is formulated. We now indicate why this occurs. Suppose $\boldsymbol{\psi} = \boldsymbol{\psi}(\boldsymbol{\theta})$ is a one-to-one transformation of the parameter vector $\boldsymbol{\theta}$ to a new p-dimensional parameter $\boldsymbol{\psi}$. For simplicity we assume $\boldsymbol{\theta}_0 = \mathbf{0}_p$, $\boldsymbol{\psi}(\mathbf{0}_p) = \mathbf{0}_p$, and that $\boldsymbol{\psi}(\boldsymbol{\theta}) = (\psi_1(\theta_1), \ldots, \psi_p(\theta_p))^{\top}$. Hence, we can rewrite the null hypothesis as $\mathcal{H}_0^{\dagger} : \boldsymbol{\psi}(\boldsymbol{\theta}) = \mathbf{0}_p$, and the matrix

$$\frac{\partial \boldsymbol{\psi}(\boldsymbol{\theta})}{\partial \boldsymbol{\theta}^{\top}} = ((\psi_r^{(s)}))$$

is diagonal, where $\psi_r^{(s)} = \psi_r^{(s)}(\theta_r) = \partial \psi_r(\theta_r)/\partial \theta_s$, for $r, s = 1, \ldots, p$. We have that

$$\boldsymbol{U}(\boldsymbol{\psi}) = \left(\frac{\partial \boldsymbol{\psi}(\boldsymbol{\theta})}{\partial \boldsymbol{\theta}^{\top}}\right)^{-1} \boldsymbol{U}(\boldsymbol{\theta}),$$

and hence the gradient statistic given by the $\boldsymbol{\psi}$-parameterization is

$$S_{\mathrm{T}}^{\dagger} = \boldsymbol{U}(\mathbf{0}_p)^{\top} \left(\frac{\partial \boldsymbol{\psi}(\boldsymbol{\theta})}{\partial \boldsymbol{\theta}^{\top}}\right)^{-1}_{\boldsymbol{\theta}=\mathbf{0}_p} \boldsymbol{\psi}(\hat{\boldsymbol{\theta}}).$$

For any number $m > 0$ and any value of $\boldsymbol{\theta}$, there exists a function $\boldsymbol{\psi}$ such that

$$\frac{\psi_r(\hat{\theta}_r)}{\psi_r^{(r)}(0)} = m\hat{\theta}_r, \quad r = 1, \ldots, p.$$

Applying this function we have that

$$\left(\frac{\partial \boldsymbol{\psi}(\boldsymbol{\theta})}{\partial \boldsymbol{\theta}^{\top}}\right)^{-1}_{\boldsymbol{\theta}=\mathbf{0}_p} \boldsymbol{\psi}(\hat{\boldsymbol{\theta}}) = \left(\frac{\psi_1(\hat{\theta}_1)}{\psi_1^{(1)}(0)}, \ldots, \frac{\psi_p(\hat{\theta}_p)}{\psi_p^{(p)}(0)}\right)^{\top}$$
$$= m\hat{\boldsymbol{\theta}},$$

and, therefore, $S_{\mathrm{T}}^{\dagger} = mS_{\mathrm{T}}$. By choosing m appropriately, it is possible to obtain any positive value of $S_{\mathrm{T}}^{\dagger}$.

Although the gradient statistic in Eq. (1.1) was derived by Terrell [6] from the score and Wald statistics, it is of a different nature. The score statistic measures the squared length of the score vector evaluated at $\mathcal{H}_0$ using the metric given by the inverse of the Fisher information matrix, whereas the Wald statistic gives the squared distance between the unrestricted and restricted MLEs of $\boldsymbol{\theta}$ using the metric given by the Fisher information matrix. Moreover, both are quadratic forms. The gradient statistic, on the other hand, is not a quadratic form and measures the distance between the unrestricted and restricted MLEs of $\boldsymbol{\theta}$ from a different perspective. It measures the orthogonal projection of the score vector evaluated at $\mathcal{H}_0$ on the vector $\hat{\boldsymbol{\theta}} - \boldsymbol{\theta}_0$. The readers are referred to Muggeo and Lovison [7] for interesting geometrical and graphical interpretations of the gradient statistic.

Some examples are presented in the following.

Example 1.1. Let $x_1, \ldots, x_n$ be n independent observations from a normal distribution with mean $\theta \in \mathbb{R}$ and with known variance, which for convenience we take as unity. Suppose the interest lies in testing $\mathcal{H}_0 : \theta = \theta_0$ against $\mathcal{H}_a : \theta \neq \theta_0$, where θ_0 is a known value. The statistic S_{T} for testing $\mathcal{H}_0 : \theta = \theta_0$ becomes

$$S_{\mathrm{T}} = n(\bar{x} - \theta_0)^2,$$

where $\bar{x} = n^{-1} \sum_{l=1}^{n} x_l$.

Example 1.2. Let $x_1, \ldots, x_n$ be a sample of size n from an exponential distribution with mean $\theta > 0$. We want to test the null hypothesis $\mathcal{H}_0 : \theta = \theta_0$ against $\mathcal{H}_a : \theta \neq \theta_0$, where θ_0 is a known positive value. We have that

$$S_{\mathrm{T}} = n\left(\frac{\bar{x} - \theta_0}{\theta_0}\right)^2,$$

where $\bar{x} = n^{-1} \sum_{l=1}^{n} x_l$.

Example 1.3. Let the counts in a set of k independent categories have observed counts n_j ($j = 1, \ldots, k$) hypothesized to follow Poisson distributions with mean θ_j. In this case, the MLE of θ_j is $\hat{\theta}_j = n_j$. It then follows that the gradient statistic reduces to

$$S_{\mathrm{T}} = \sum_{j=1}^{k} \frac{(n_j - \theta_j)^2}{\theta_j};$$

that is, the Pearson goodness-of-fit statistic.

Example 1.4. Let (x_l, y_l), $l = 1, \ldots, n$, be n independent and identically distributed random variables with common probability density function

$$f(x, y; \theta) = \exp\left(-\theta x - \frac{y}{\theta}\right), \quad x, y > 0, \quad \theta > 0,$$

which is usually referred to as Fisher's gamma hyperbola. The log-likelihood function is $\ell(\theta) = -n(\theta\bar{x} + \bar{y}/\theta)$, where $\bar{x} = n^{-1}\sum_{l=1}^{n} x_l$ and $\bar{y} = n^{-1}\sum_{l=1}^{n} y_l$, and the MLE of θ is given by $\hat{\theta} = (\bar{y}/\bar{x})^{1/2}$. The gradient statistic to test $\mathcal{H}_0 : \theta = \theta_0$, where θ_0 is a known positive value, takes the form

$$S_{\mathrm{T}} = n\left(\frac{\bar{y} - \theta_0^2\bar{x}}{\theta_0^2}\right)\left(\frac{\bar{y}^{1/2} - \theta_0\bar{x}^{1/2}}{\bar{x}^{1/2}}\right).$$

1.3 SOME PROPERTIES OF THE GRADIENT STATISTIC

Firstly, the asymptotic distribution of the gradient statistic S_{T} is provided in the following theorem.

Theorem 1.1. *Under the null hypothesis $\mathcal{H}_0 : \boldsymbol{\theta} = \boldsymbol{\theta}_0$, the gradient statistic $S_{\mathrm{T}} = \boldsymbol{U}(\boldsymbol{\theta}_0)^{\top}(\hat{\boldsymbol{\theta}} - \boldsymbol{\theta}_0)$ has a central χ_p^2 distribution up to an error of order $O(n^{-1})$:*

$$\Pr(S_{\mathrm{T}} \leq x|\mathcal{H}_0) = G_p(x) + O(n^{-1}),$$

where $G_p(\cdot)$ denotes the central χ^2 cumulative distribution function with p degrees of freedom.

Proof. By a Taylor expansion of $\boldsymbol{U}(\hat{\boldsymbol{\theta}})$ about $\boldsymbol{\theta}_0$ and under mild regularity conditions, it can be shown that

$$\hat{\boldsymbol{\theta}} - \boldsymbol{\theta}_0 = \boldsymbol{K}(\boldsymbol{\theta}_0)^{-1}\boldsymbol{U}(\boldsymbol{\theta}_0) + O_p(n^{-1}).$$

Since $\boldsymbol{U}(\boldsymbol{\theta}) = O_p(n^{1/2})$, we have that

$$\begin{aligned}\boldsymbol{U}(\boldsymbol{\theta}_0)^\top(\hat{\boldsymbol{\theta}} - \boldsymbol{\theta}_0) &= \boldsymbol{U}(\boldsymbol{\theta}_0)^\top[\boldsymbol{K}(\boldsymbol{\theta}_0)^{-1}\boldsymbol{U}(\boldsymbol{\theta}_0) + O_p(n^{-1})]\\ &= \boldsymbol{U}(\boldsymbol{\theta}_0)^\top\boldsymbol{K}(\boldsymbol{\theta}_0)^{-1}\boldsymbol{U}(\boldsymbol{\theta}_0) + O_p(n^{-1/2});\end{aligned}$$

that is, $S_{\mathrm{T}} = S_{\mathrm{R}} + O_p(n^{-1/2}) = S_{\mathrm{R}} + o_p(1)$. Since the statistic S_{R} has an asymptotic central χ^2_p distribution under $\mathcal{H}_0$, the result holds. □

Theorem 1.1 reveals that the gradient, LR, Wald, and score statistics have the same asymptotic distribution under the null hypothesis. It also indicates that we can reject the null hypothesis $\mathcal{H}_0 : \boldsymbol{\theta} = \boldsymbol{\theta}_0$ if the observed value of the gradient statistic S_{T} exceeds the upper $100(1-\gamma)\%$ quantile of the χ^2_p distribution for a given nominal level.

As pointed out in Terrell [6], the gradient statistic has one peculiarity: it is not transparently nonnegative, even though it must be so asymptotically. Since the LR, Wald, and score statistics are obviously nonnegative, the question is natural. Terrell [6, p. 207] wrote "at the seminar at which the statistic was introduced, this was the first question from the floor." The next theorem deals with this peculiarity.

Theorem 1.2. *If the log-likelihood function $\ell(\boldsymbol{\theta})$ is unimodal and differentiable at some $\boldsymbol{\theta}_0 \in \boldsymbol{\Theta}$, then*

$$S_{\mathrm{T}} = \boldsymbol{U}(\boldsymbol{\theta}_0)^\top(\hat{\boldsymbol{\theta}} - \boldsymbol{\theta}_0) \geq 0.$$

Proof. Readers are referred to Terrell [6, p. 208]. □

The next theorem provides an interesting (and curious as well) property of the gradient statistic.

Theorem 1.3. *Let $\bar{\boldsymbol{\theta}}$ be an unbiased estimator for $\boldsymbol{\theta} \in \boldsymbol{\Theta}$. Then,*

$$\mathbb{E}[\boldsymbol{U}(\boldsymbol{\theta})^\top(\bar{\boldsymbol{\theta}} - \boldsymbol{\theta})] = p,$$

for all $\boldsymbol{\theta} \in \boldsymbol{\Theta}$.

Proof. Let $\bar{\boldsymbol{\theta}}$ be indeed an unbiased estimator for $\boldsymbol{\theta}$: $\mathbb{E}(\bar{\boldsymbol{\theta}}) = \boldsymbol{\theta}$ for all $\boldsymbol{\theta} \in \boldsymbol{\Theta}$. Note that

$$\mathbb{COV}(\bar{\boldsymbol{\theta}}, \boldsymbol{U}(\boldsymbol{\theta})) = \int \bar{\boldsymbol{\theta}} \boldsymbol{U}(\boldsymbol{\theta}) Lik(\boldsymbol{\theta})\, d\boldsymbol{x} = \int \bar{\boldsymbol{\theta}} \frac{\partial \ell(\boldsymbol{\theta})}{\partial \boldsymbol{\theta}^{\top}} Lik(\boldsymbol{\theta})\, d\boldsymbol{x}$$
$$= \int \frac{\bar{\boldsymbol{\theta}}}{Lik(\boldsymbol{\theta})} \frac{\partial Lik(\boldsymbol{\theta})}{\partial \boldsymbol{\theta}^{\top}} Lik(\boldsymbol{\theta})\, d\boldsymbol{x} = \int \bar{\boldsymbol{\theta}} \frac{\partial Lik(\boldsymbol{\theta})}{\partial \boldsymbol{\theta}^{\top}}\, d\boldsymbol{x},$$

where $Lik(\boldsymbol{\theta})$ is the likelihood function, and $\boldsymbol{x} = (x_1, \ldots, x_n)^{\top}$. By inverting the order of the integration and differentiation, it follows that

$$\int \bar{\boldsymbol{\theta}} \frac{\partial Lik(\boldsymbol{\theta})}{\partial \boldsymbol{\theta}^{\top}}\, d\boldsymbol{x} = \frac{\partial}{\partial \boldsymbol{\theta}^{\top}} \int \bar{\boldsymbol{\theta}} Lik(\boldsymbol{\theta})\, d\boldsymbol{x}.$$

By using the fact that $\bar{\boldsymbol{\theta}}$ is an unbiased estimator for $\boldsymbol{\theta}$, we have that

$$\frac{\partial}{\partial \boldsymbol{\theta}^{\top}} \int \bar{\boldsymbol{\theta}} Lik(\boldsymbol{\theta})\, d\boldsymbol{x} = \frac{\partial \boldsymbol{\theta}}{\partial \boldsymbol{\theta}^{\top}} = \boldsymbol{I}_p.$$

Hence, note that

$$\begin{aligned}
\mathbb{E}\left[\boldsymbol{U}(\boldsymbol{\theta})^{\top}\left(\bar{\boldsymbol{\theta}} - \boldsymbol{\theta}\right)\right] &= \mathbb{E}\left[\operatorname{tr}\left\{\boldsymbol{U}(\boldsymbol{\theta})^{\top}\left(\bar{\boldsymbol{\theta}} - \boldsymbol{\theta}\right)\right\}\right] \\
&= \mathbb{E}\left[\operatorname{tr}\left\{\left(\bar{\boldsymbol{\theta}} - \boldsymbol{\theta}\right)\boldsymbol{U}(\boldsymbol{\theta})^{\top}\right\}\right] \\
&= \operatorname{tr}\left\{\mathbb{E}\left[\left(\bar{\boldsymbol{\theta}} - \boldsymbol{\theta}\right)\boldsymbol{U}(\boldsymbol{\theta})^{\top}\right]\right\} \\
&= \operatorname{tr}\left\{\mathbb{COV}\left(\bar{\boldsymbol{\theta}}, \boldsymbol{U}(\boldsymbol{\theta})\right)\right\} \\
&= \operatorname{tr}\left\{\boldsymbol{I}_p\right\} \\
&= p,
\end{aligned}$$

where $\operatorname{tr}\{\cdot\}$ denotes the trace of a matrix. □

Theorem 1.3 points out an interesting feature of the gradient statistic. According to Terrell [6], it suggests that we may often improve the χ^2 approximation of the gradient statistic in Eq. (1.1) by using a less-biased estimate for $\boldsymbol{\theta}$ in the place of $\hat{\boldsymbol{\theta}}$. This may be accomplished in two ways. As noted earlier, the gradient statistic lacks one desirable feature of the LR and score statistics: it is not invariant under nonlinear reparameterization of $\boldsymbol{\theta}$. However, it means that we may improve its behavior by choosing a parameterization in which the MLE of $\boldsymbol{\theta}$ is unbiased. Another way to take advantage of Theorem 1.3 is to adjust the MLE to a new asymptotically efficient estimate $\bar{\boldsymbol{\theta}}$ that is unbiased or nearly so and thereby improve the χ^2 approximation. It is worth emphasizing that the nonnegativity of the gradient statistic is not guaranteed when we use an estimator for $\boldsymbol{\theta}$ in Eq. (1.1) that is not the MLE.

Some Monte Carlo simulation experiments presented in Lemonte and Ferrari [8] for testing hypotheses on the parameters of the Birnbaum-Saunders distribution under type II censored samples have indicated that the χ^2 approximation for the null distribution of the gradient statistic in fact improves when a new estimator which is unbiased (or nearly so) is considered in the place of the MLE. To avoid possible negative values for S_{T} when the MLE is replaced by a less-biased estimate, these authors have defined a modified gradient statistic; see Section 1.5 for more details.

1.4 COMPOSITE NULL HYPOTHESIS

Let the parameter vector $\boldsymbol{\theta}$ be partitioned as $\boldsymbol{\theta} = (\boldsymbol{\theta}_1^\top, \boldsymbol{\theta}_2^\top)^\top$, where $\boldsymbol{\theta}_1$ and $\boldsymbol{\theta}_2$ are parameter vectors of dimensions q and $p - q$, respectively. Suppose the interest lies in testing the composite null hypothesis $\mathcal{H}_0 : \boldsymbol{\theta}_1 = \boldsymbol{\theta}_{10}$ against the two-sided alternative hypothesis $\mathcal{H}_a : \boldsymbol{\theta}_1 \neq \boldsymbol{\theta}_{10}$, where $\boldsymbol{\theta}_{10}$ is a fixed q-dimensional vector. Hence, $\boldsymbol{\theta}_2$ is a $(p - q)$-vector of nuisance parameters. Let $\hat{\boldsymbol{\theta}} = (\hat{\boldsymbol{\theta}}_1^\top, \hat{\boldsymbol{\theta}}_2^\top)^\top$ and $\tilde{\boldsymbol{\theta}} = (\boldsymbol{\theta}_{10}^\top, \tilde{\boldsymbol{\theta}}_2^\top)^\top$ be the unrestricted and restricted (obtained under the null hypothesis) MLEs of $\boldsymbol{\theta} = (\boldsymbol{\theta}_1^\top, \boldsymbol{\theta}_2^\top)^\top$, respectively. The gradient statistic for testing the composite null hypothesis $\mathcal{H}_0 : \boldsymbol{\theta}_1 = \boldsymbol{\theta}_{10}$ is defined as

$$S_{\mathrm{T}} = \boldsymbol{U}(\tilde{\boldsymbol{\theta}})^\top (\hat{\boldsymbol{\theta}} - \tilde{\boldsymbol{\theta}}). \tag{1.2}$$

The gradient statistic in Eq. (1.2) has an asymptotic χ_q^2 distribution under the null hypothesis, q being the number of restrictions imposed by $\mathcal{H}_0$. The null hypothesis $\mathcal{H}_0 : \boldsymbol{\theta}_1 = \boldsymbol{\theta}_{10}$ is rejected if the observed value of the statistic S_{T} exceeds the upper $100(1 - \gamma)\%$ quantile of the χ_q^2 distribution for a given nominal level.

The partition for $\boldsymbol{\theta}$ induces the corresponding partition: $\boldsymbol{U}(\boldsymbol{\theta}) = (\boldsymbol{U}_1(\boldsymbol{\theta})^\top, \boldsymbol{U}_2(\boldsymbol{\theta})^\top)^\top$, where

$$\boldsymbol{U}_1(\boldsymbol{\theta}) = \frac{\partial \ell(\boldsymbol{\theta})}{\partial \boldsymbol{\theta}_1}, \quad \boldsymbol{U}_2(\boldsymbol{\theta}) = \frac{\partial \ell(\boldsymbol{\theta})}{\partial \boldsymbol{\theta}_2}.$$

Since $\boldsymbol{U}_2(\tilde{\boldsymbol{\theta}}) = \mathbf{0}_{p-q}$, the gradient statistic in Eq. (1.2) can be expressed as

$$S_{\mathrm{T}} = \boldsymbol{U}_1(\tilde{\boldsymbol{\theta}})^\top (\hat{\boldsymbol{\theta}}_1 - \boldsymbol{\theta}_{10}). \tag{1.3}$$

The above expression is very convenient for computing purposes. In the following we present some examples.

Example 1.5. Let $x_1, \ldots, x_n$ be n independent observations from a Weibull distribution with shape parameter $\beta > 0$, and scale parameter $\alpha > 0$. The Weibull probability density function is given by $f(x; \alpha, \beta) = (\beta/\alpha^{\beta})x^{\beta-1}\exp[-(x/\alpha)^{\beta}]$, $x > 0$. We wish to test the null hypothesis $\mathcal{H}_0 : \beta = 1$, which means that the data come from an exponential distribution with scale parameter α, against $\mathcal{H}_{\text{a}} : \beta \neq 1$. Here, α acts as a nuisance parameter. The gradient statistic can be written in the form

$$S_{\text{T}} = n(\hat{\beta} - 1)\left(1 + \bar{x}_1 - \frac{\bar{x}_2}{\bar{x}}\right),$$

where $\bar{x} = n^{-1}\sum_{l=1}^{n} x_l$, $\bar{x}_1 = n^{-1}\sum_{l=1}^{n} \log x_l$, $\bar{x}_2 = n^{-1}\sum_{l=1}^{n} x_l \log x_l$, and $\hat{\beta}$ is the MLE of β, which can be obtained implicitly from the equation

$$\hat{\beta}^{-1} = \frac{\sum_{l=1}^{n} x_l^{\hat{\beta}} \log x_l}{\sum_{l=1}^{n} x_l^{\hat{\beta}}} - \bar{x}_1.$$

Under the null hypothesis, the statistic S_{T} has a limiting χ_1^2 distribution.

Example 1.6. Let $x_{11}, \ldots, x_{1n_1}$ and $x_{21}, \ldots, x_{2n_2}$ be two independent samples from exponential distributions with means λ and $\psi\lambda$, respectively. The parameter of interest is ψ — the ratio of the means — and the interest lies in testing $\mathcal{H}_0 : \psi = 1$, which is equivalent to the equality of the two population means, against $\mathcal{H}_{\text{a}} : \psi \neq 1$. We consider the balanced case ($n_1 = n_2 = n/2$, $n \geq 2$ even). The gradient statistic for testing $\mathcal{H}_0$ takes the form

$$S_{\text{T}} = \frac{n(\bar{x}_1 - \bar{x}_2)^2}{4\bar{x}_1\bar{x}},$$

where $\bar{x}_1 = n_1^{-1}\sum_{l=1}^{n_1} x_{1l}$, $\bar{x}_2 = n_2^{-1}\sum_{l=1}^{n_2} x_{2l}$, and $\bar{x} = (\bar{x}_1 + \bar{x}_2)/2$. Under the null hypothesis, S_{T} has a limiting χ_1^2 distribution.

Example 1.7. The generalized inverse Gaussian (GIG) distribution is widely used for modeling and analyzing lifetime data. A random variable X has a GIG distribution if its probability density function is given by

$$f(x; \lambda, \chi, \psi) = \frac{(\psi/\chi)^{\lambda/2}}{2B_{\lambda}(\sqrt{\chi\psi})} x^{\lambda-1} \exp\left[-\frac{1}{2}\left(\psi x + \frac{\chi}{x}\right)\right], \quad x > 0.$$

Here, $\lambda \in \mathbb{R}$, $(\chi, \psi) \in \boldsymbol{\Theta}_\lambda$, where $\boldsymbol{\Theta}_\lambda = \{(\chi, \psi) : \chi \geq 0, \psi > 0\}$ if $\lambda > 0$, $\boldsymbol{\Theta}_\lambda = \{(\chi, \psi) : \chi > 0, \psi > 0\}$ if $\lambda = 0$, and $\boldsymbol{\Theta}_\lambda = \{(\chi, \psi) : \chi > 0, \psi \geq 0\}$ if $\lambda < 0$. Also, $B_\nu(z)$ denotes the modified Bessel function of the third kind with index ν and argument z. Its integral representation is $B_\nu(z) = (1/2)\int_{-\infty}^{\infty} \exp[-z\cosh(t) - \nu t]\, dt$. Special models include the gamma distribution ($\chi = 0, \lambda > 0$), the reciprocal gamma distribution ($\psi = 0, \lambda < 0$), the inverse Gaussian distribution ($\lambda = -1/2$), and the hyperbola distribution ($\lambda = 0$). Introducing the parameters $\omega = \chi/2$ and $\eta = \psi/2$, the above density function becomes

$$f(x; \lambda, \omega, \eta) = Cx^{\lambda-1} \exp\left[-\left(\eta x + \omega x^{-1}\right)\right], \quad x > 0,$$

where C is the normalizing constant given by $C = C(\lambda, \omega, \eta) = (\eta/\omega)^{\lambda/2} / [2B_\lambda(2\sqrt{\eta\omega})]$. Let $R_\lambda = R_\lambda(\eta, \omega) = B_{\lambda+1}(2\sqrt{\eta\omega})/B_\lambda(2\sqrt{\eta\omega})$. Let $x_1, \ldots, x_n$ denote n independent observations from the GIG distribution. First, suppose the interest lies in testing the null hypothesis $\mathcal{H}_0 : \eta = \eta_0$ against $\mathcal{H}_\mathrm{a} : \eta \neq \eta_0$, where η_0 is a known positive value, and λ and ω are nuisance parameters. The gradient statistic for testing $\mathcal{H}_0 : \eta = \eta_0$ is given by

$$S_\mathrm{T} = n(\hat{\eta} - \eta_0)\left(\frac{\tilde{\omega}^{1/2}\tilde{R}_\lambda}{\eta_0^{1/2}} - \bar{x}\right),$$

where $\bar{x} = n^{-1}\sum_{l=1}^{n} x_l$, $\hat{\eta}$ is the MLE of η, $\tilde{R}_\lambda = R_{\tilde{\lambda}}(\eta_0, \tilde{\omega})$, and $\tilde{\lambda}$ and $\tilde{\omega}$ are the restricted (obtained under $\mathcal{H}_0 : \eta = \eta_0$) MLEs of λ and ω, respectively. Now, we wish to test $\mathcal{H}_0 : \omega = \omega_0$ against $\mathcal{H}_\mathrm{a} : \omega \neq \omega_0$, where ω_0 is a known positive value, and λ and η act as nuisance parameters. The gradient statistic for testing $\mathcal{H}_0 : \omega = \omega_0$ takes the form

$$S_\mathrm{T} = n(\hat{\omega} - \omega_0)\left(-\frac{\tilde{\lambda}}{\omega_0} + \frac{\tilde{\eta}^{1/2}\tilde{R}_\lambda}{\omega_0^{1/2}} - \bar{h}\right),$$

where $\bar{h} = n^{-1}\sum_{l=1}^{n} x_l^{-1}$, $\hat{\omega}$ is the MLE of ω, $\tilde{R}_\lambda = R_{\tilde{\lambda}}(\tilde{\eta}, \omega_0)$, and $\tilde{\lambda}$ and $\tilde{\eta}$ are the restricted (obtained under $\mathcal{H}_0 : \omega = \omega_0$) MLEs of λ and η, respectively. Under both null hypotheses, the gradient statistic has a limiting χ_1^2 distribution.

Example 1.8. Let

$$\begin{pmatrix} x_l \\ y_l \end{pmatrix} \sim \mathcal{N}_2\left[\begin{pmatrix} \mu_x \\ \mu_y \end{pmatrix}, \begin{pmatrix} \sigma_x^2 & \rho\sigma_x\sigma_y \\ \rho\sigma_x\sigma_y & \sigma_y^2 \end{pmatrix}\right], \quad l = 1, \ldots, n,$$

where $\mu_x \in \mathbb{R}$, $\mu_y \in \mathbb{R}$, $\sigma_x^2 > 0$, $\sigma_y^2 > 0$, and $\rho \in (-1, 1)$ is the correlation coefficient. The MLEs of μ_x, μ_y, σ_x^2, σ_y^2, and ρ are

$$\hat{\mu}_x = \bar{x}, \quad \hat{\mu}_y = \bar{y}, \quad \hat{\sigma}_x^2 = \frac{1}{n}\sum_{l=1}^{n}(x_l - \bar{x})^2, \quad \hat{\sigma}_y^2 = \frac{1}{n}\sum_{l=1}^{n}(y_l - \bar{y})^2,$$

$$\hat{\rho} = \frac{\sum_{l=1}^{n}(x_l - \bar{x})(y_l - \bar{y})}{\left[\sum_{l=1}^{n}(x_l - \bar{x})^2 \sum_{l=1}^{n}(y_l - \bar{y})^2\right]^{1/2}},$$

where $\bar{x} = n^{-1}\sum_{l=1}^{n} x_l$ and $\bar{y} = n^{-1}\sum_{l=1}^{n} y_l$. Suppose the interest lies in testing $\mathcal{H}_0 : \rho = 0$ against $\mathcal{H}_a : \rho \neq 0$, where the null hypothesis means independency. The gradient statistic to test $\mathcal{H}_0 : \rho = 0$ can be reduced simply to

$$S_T = n\left(\frac{\sum_{l=1}^{n}(x_l - \bar{x})(y_l - \bar{y})}{\left[\sum_{l=1}^{n}(x_l - \bar{x})^2 \sum_{l=1}^{n}(y_l - \bar{y})^2\right]^{1/2}}\right)^2 = n\hat{\rho}^2,$$

which follows asymptotically a χ_1^2 distribution under the null hypothesis.

Example 1.9. Suppose x_{ij} are independent random variables with distributions $\mathcal{N}(\mu_i, \sigma_i^2)$, for $i = 1, \ldots, g$ and $j = 1, \ldots, n_i$. We want to test the null hypothesis of homogeneity of variances among the g groups

$$\mathcal{H}_0 : \sigma_1^2 = \cdots = \sigma_g^2,$$

against the alternative that at least one equality does not hold. The null hypothesis can be expressed as a linear constraint in the form $\mathcal{H}_0 : \psi_1 = \cdots = \psi_{g-1} = 0$, with, for instance, $\psi_i = \sigma_{i+1}^2 - \sigma_i^2$, for $i = 1, \ldots, g-1$. The log-likelihood function for the parameter vector $\boldsymbol{\theta} = (\boldsymbol{\mu}^\top, \boldsymbol{\sigma}^{2\top})^\top$, where $\boldsymbol{\mu} = (\mu_1, \ldots, \mu_g)^\top$ and $\boldsymbol{\sigma}^2 = (\sigma_1^2, \ldots, \sigma_g^2)^\top$, is given by

$$\ell(\boldsymbol{\theta}) = -\frac{1}{2}\sum_{i=1}^{g}\left[n_i \log \sigma_i^2 + \frac{1}{\sigma_i^2}\sum_{j=1}^{n_i}(x_{ij} - \mu_i)^2\right].$$

Note that

$$(\sigma_i^2)' = \frac{\partial \ell(\boldsymbol{\theta})}{\partial \sigma_i^2} = -\frac{n_i}{2\sigma_i^2} + \frac{1}{2(\sigma_i^2)^2}\sum_{j=1}^{n_i}(x_{ij} - \mu_i)^2, \quad i = 1, \ldots, g,$$

and hence $\boldsymbol{U}_{\boldsymbol{\sigma}^2}(\boldsymbol{\theta}) = \partial \ell(\boldsymbol{\theta})/\partial \boldsymbol{\sigma}^2 = ((\sigma_1^2)', \ldots, (\sigma_g^2)')^\top$. We can show that the unrestricted and restricted MLEs of $\boldsymbol{\theta} = (\boldsymbol{\mu}^\top, \boldsymbol{\sigma}^{2\top})^\top$ reduce to $\hat{\boldsymbol{\theta}} = (\hat{\boldsymbol{\mu}}^\top, \hat{\boldsymbol{\sigma}}^{2\top})^\top$ and $\tilde{\boldsymbol{\theta}} = (\hat{\boldsymbol{\mu}}^\top, \tilde{\boldsymbol{\sigma}}^{2\top})^\top$, respectively, where $\hat{\boldsymbol{\mu}} = (\bar{x}_1, \ldots, \bar{x}_g)^\top$,

$\hat{\boldsymbol{\sigma}}^2 = (\hat{\sigma}_1^2, \ldots, \hat{\sigma}_g^2)^\top$, and $\tilde{\boldsymbol{\sigma}}^2 = (\tilde{\sigma}^2, \ldots, \tilde{\sigma}^2)^\top$. Here, $\bar{x}_i = n_i^{-1} \sum_{j=1}^{n_i} x_{ij}$, $\hat{\sigma}_i^2 = n_i^{-1} \sum_{j=1}^{n_i} (x_{ij} - \bar{x}_i)^2$, and $\tilde{\sigma}^2 = \sum_{i=1}^{g} n_i \hat{\sigma}_i^2 / \sum_{i=1}^{g} n_i$. The gradient statistic to test the null hypothesis $\mathcal{H}_0 : \sigma_1^2 = \cdots = \sigma_g^2$ is given by $S_{\mathrm{T}} = \boldsymbol{U}_{\sigma^2}(\tilde{\boldsymbol{\theta}})^\top (\hat{\boldsymbol{\sigma}}^2 - \tilde{\boldsymbol{\sigma}}^2)$ and takes the form

$$S_{\mathrm{T}} = \frac{1}{2\tilde{\sigma}^2} \sum_{i=1}^{g} \left[n_i(\tilde{\sigma}^2 - \hat{\sigma}_i^2) + \frac{(\hat{\sigma}_i^2 - \tilde{\sigma}^2)}{\tilde{\sigma}^2} \sum_{j=1}^{n_i} (x_{ij} - \bar{x}_i)^2 \right],$$

which follows asymptotically the χ_{g-1}^2 distribution under the null hypothesis.

Example 1.10. Let $\boldsymbol{y} = \boldsymbol{X}\boldsymbol{\beta} + \boldsymbol{u}$ with $\boldsymbol{u} \sim \mathcal{N}_n(\boldsymbol{0}_n, \boldsymbol{\Omega})$, where $\boldsymbol{y}$ is $n \times 1$, $\boldsymbol{X}$ is $n \times p$, $\boldsymbol{\beta}$ is $p \times 1$, and $\boldsymbol{\Omega}$ is $n \times n$ and known. Suppose the interest lies in testing the null hypothesis $\mathcal{H}_0 : \boldsymbol{R}\boldsymbol{\beta} = \boldsymbol{r}$ against the alternative hypothesis $\mathcal{H}_{\mathrm{a}}$: $\boldsymbol{R}\boldsymbol{\beta} \neq \boldsymbol{r}$, where $\boldsymbol{R}$ is $q \times p$ and $\boldsymbol{r}$ is $q \times 1$, both of which are known. The score function is given by $\boldsymbol{U}(\boldsymbol{\beta}) = \boldsymbol{X}^\top \boldsymbol{\Omega}^{-1}(\boldsymbol{y} - \boldsymbol{X}\boldsymbol{\beta})$. The unrestricted and restricted MLEs of $\boldsymbol{\beta}$ are $\hat{\boldsymbol{\beta}} = \boldsymbol{C}^{-1}\boldsymbol{X}^\top \boldsymbol{\Omega}^{-1}\boldsymbol{y}$ and $\tilde{\boldsymbol{\beta}} = \hat{\boldsymbol{\beta}} - \boldsymbol{C}^{-1}\boldsymbol{R}^\top(\boldsymbol{R}\boldsymbol{C}^{-1}\boldsymbol{R}^\top)^{-1} (\boldsymbol{R}\hat{\boldsymbol{\beta}} - \boldsymbol{r})$, respectively, where $\boldsymbol{C} = \boldsymbol{X}^\top \boldsymbol{\Omega}^{-1}\boldsymbol{X}$. Notice that $\boldsymbol{U}(\tilde{\boldsymbol{\beta}}) = (\boldsymbol{R}\boldsymbol{C}^{-1}\boldsymbol{R}^\top)^{-1}(\boldsymbol{R}\hat{\boldsymbol{\beta}} - \boldsymbol{r})$ and hence the gradient statistic becomes

$$S_{\mathrm{T}} = (\boldsymbol{R}\hat{\boldsymbol{\beta}} - \boldsymbol{r})^\top (\boldsymbol{R}\boldsymbol{C}^{-1}\boldsymbol{R}^\top)^{-1}(\boldsymbol{R}\hat{\boldsymbol{\beta}} - \boldsymbol{r}),$$

which has an asymptotic χ_q^2 distribution under $\mathcal{H}_0$.

In Example (1.10), we have that $\mathbb{E}(\hat{\boldsymbol{\beta}}) = \boldsymbol{\beta}$ and hence $\hat{\boldsymbol{\beta}}$ is an unbiased estimator for $\boldsymbol{\beta}$. Note that

$$\begin{aligned}
\mathbb{E}(S_{\mathrm{T}}) &= \mathbb{E}[(\boldsymbol{R}\hat{\boldsymbol{\beta}} - \boldsymbol{r})^\top (\boldsymbol{R}\boldsymbol{C}^{-1}\boldsymbol{R}^\top)^{-1}(\boldsymbol{R}\hat{\boldsymbol{\beta}} - \boldsymbol{r})] \\
&= \mathbb{E}[\mathrm{tr}\{(\boldsymbol{R}\hat{\boldsymbol{\beta}} - \boldsymbol{r})^\top (\boldsymbol{R}\boldsymbol{C}^{-1}\boldsymbol{R}^\top)^{-1}(\boldsymbol{R}\hat{\boldsymbol{\beta}} - \boldsymbol{r})\}] \\
&= \mathbb{E}[\mathrm{tr}\{(\boldsymbol{R}\hat{\boldsymbol{\beta}} - \boldsymbol{r})(\boldsymbol{R}\hat{\boldsymbol{\beta}} - \boldsymbol{r})^\top (\boldsymbol{R}\boldsymbol{C}^{-1}\boldsymbol{R}^\top)^{-1}\}] \\
&= \mathrm{tr}\{\mathbb{E}[(\boldsymbol{R}\hat{\boldsymbol{\beta}} - \boldsymbol{r})(\boldsymbol{R}\hat{\boldsymbol{\beta}} - \boldsymbol{r})^\top](\boldsymbol{R}\boldsymbol{C}^{-1}\boldsymbol{R}^\top)^{-1}\}.
\end{aligned}$$

By noting that $\mathbb{E}[(\boldsymbol{R}\hat{\boldsymbol{\beta}} - \boldsymbol{r})(\boldsymbol{R}\hat{\boldsymbol{\beta}} - \boldsymbol{r})^\top] = \boldsymbol{R}\boldsymbol{C}^{-1}\boldsymbol{R}^\top$, since $\mathbb{COV}(\hat{\boldsymbol{\beta}}) = (\boldsymbol{X}^\top \boldsymbol{\Omega}^{-1}\boldsymbol{X})^{-1}$, it follows that

$$\begin{aligned}
\mathbb{E}(S_{\mathrm{T}}) &= \mathrm{tr}\{\boldsymbol{R}\boldsymbol{C}^{-1}\boldsymbol{R}^\top (\boldsymbol{R}\boldsymbol{C}^{-1}\boldsymbol{R}^\top)^{-1}\} \\
&= \mathrm{tr}\{\boldsymbol{I}_q\} \\
&= q,
\end{aligned}$$

Table 1.1 The Mandible Lengths (in mm) of 10 Female and 10 Male Jackal Skulls

Length	Sex	Length	Sex	Length	Sex	Length	Sex
105	F	108	F	111	M	114	M
106	F	110	F	111	F	114	M
107	M	110	M	111	F	116	M
107	F	110	F	112	M	117	M
107	F	111	F	113	M	120	M

which is exactly the number of restrictions imposed in the null hypothesis $\mathcal{H}_0 : \boldsymbol{R\beta} = \boldsymbol{r}$. It illustrates a simple application of Theorem 1.3.

The next example deals with a simple numerical computation of the gradient statistic. It was taken from Terrell [6]. The data used in the next example are listed in Table 1.1.

Example 1.11. We consider a linear logistic regression model for the data in Table 1.1. Suppose $Y_l = \{0 : \textit{female}; 1 : \textit{male}\}$ has Bernoulli distribution with probability of success p_l, $Y_l \sim \text{Bernoulli}(p_l)$, for $l = 1, \ldots, 20$. We consider the model

$$\log\left(\frac{p_l}{1 - p_l}\right) = \beta_1 + \beta_2(x_l - \bar{x}), \quad l = 1, \ldots, 20,$$

where $\bar{x} = (1/20)\sum_{l=1}^{20} x_l = 111$, and x_l is the lth mandible length. The MLEs of the model parameters (asymptotic standard errors between parentheses) are $\hat{\beta}_1 = 0.1507\,(0.6033)$ and $\hat{\beta}_2 = 0.6085\,(0.2812)$. So, can we predict the sex of a jackal from the mandible length of a skull? The null hypothesis we wish to test is $\mathcal{H}_0 : \beta_2 = 0$. In this case, the restricted MLE of β_1 is approximately zero and the gradient statistic is computed as

$$S_{\mathrm{T}} = 0.6085 \sum_{l=1}^{20} x_l(\hat{p}_l - 0.5) = 14.6038,$$

where

$$\hat{p}_l = \frac{\exp[0.1507 + 0.6085(x_l - 111)]}{1 + \exp[0.1507 + 0.6085(x_l - 111)]}.$$

Therefore, based on the observed value of the gradient statistic, we reject the null hypothesis under common critical values and hence the mandible length provides a clue to sex.

At this moment, a natural question that arises is: When should we consider the gradient statistic in preference to the classical test statistics (ie, the Holy Trinity)? As remarked before, the gradient statistic does not require one to obtain, estimate, or invert an information matrix (neither expected nor observed), unlike the Wald and Rao score statistics. Also, its formal simplicity is always an attraction. As pointed out by Terrell [6], the LR statistic is even simpler in appearance than the gradient statistic, however, in many instances, the first derivative of the log-likelihood (ie, the score function) is quite a bit simpler than the log-likelihood itself. In the case of generalized linear models for ranks, for example, these derivatives are within the reach of modern computing equipment, while the likelihood itself is currently computationally intractable. In complex problems (like problems involving censored observations), where is not possible to compute, in general, the (expected) Fisher information matrix in closed-form, the gradient statistic can be used without any problem (like the LR statistic as well) and hence it represents a great advantage of this statistic in relation to its progenitors. Further, according to Terrell [6], in some particular applications, such as linear regression with Laplace errors, to require that the likelihood be twice differentiable may be too strong and hence may have problems in computing the Fisher information matrix, which implies that the Wald and score statistics may not work in such a case. Obviously the gradient statistic can be used in the regression model with Laplace errors without restrictions.

Finally, it should be emphasized that we are not recommending that users no longer consider the Wald and score statistics in their works. These two statistics have their merits and of course users may always consider them in their theoretical and applied problems. The main message in this chapter is that there is a new (and very simple as well) statistic that can be an interesting alternative to the Holy Trinity (ie, the LR, Wald, and score statistics). Therefore, we strongly recommend you give the new gradient statistic a try in theoretical and applied works.

1.5 BIRNBAUM-SAUNDERS DISTRIBUTION UNDER TYPE II CENSORING

Motivated by problems of vibration in commercial aircraft that caused fatigue in the materials, Birnbaum and Saunders [9] pioneered a two-parameter distribution to model failure time due to fatigue under cyclic loading and the assumption that failure follows from the development and growth of a dominant crack. This distribution is known as the two-parameter Birnbaum-Saunders (BS) distribution or as the fatigue life distribution. The BS distribution is an attractive alternative to the Weibull, gamma, and log-normal models, since its derivation considers the basic characteristics of the fatigue process. Additionally, it has received significant attention over the last few years by many researchers and there has been much theoretical developments with respect to this distribution. Although the BS distribution has its genesis from engineering, it has also received wide ranging applications in other fields that include business, environment, informatics, and medicine.

The cumulative distribution function of T having the BS distribution, say $T \sim \mathrm{BS}(\alpha, \eta)$, is defined by

$$F(t; \alpha, \eta) = \Phi(v_t), \quad t > 0,$$

where $\Phi(\cdot)$ is the standard normal cumulative distribution function, $v_t = \rho(t/\eta)/\alpha$, $\rho(z) = z^{1/2} - z^{-1/2}$, and $\alpha > 0$ and $\eta > 0$ are the shape and scale parameters, respectively. The scale parameter is also the median of the distribution. For any $k > 0$, it follows that $kT \sim \mathrm{BS}(\alpha, k\eta)$. It is noteworthy that the reciprocal property holds for the BS distribution: $T^{-1} \sim \mathrm{BS}(\alpha, \eta^{-1})$. The probability density function of T is

$$f(t; \alpha, \eta) = \kappa(\alpha, \eta) t^{-3/2}(t + \eta) \exp\left[-\frac{\tau(t/\eta)}{2\alpha^2}\right], \quad t > 0,$$

where $\kappa(\alpha, \eta) = \exp(\alpha^{-2})/(2\alpha\sqrt{2\pi\eta})$ and $\tau(z) = z + z^{-1}$. A general expression for the moments of T is

$$\mathbb{E}(T^k) = \eta^k \left[\frac{B_{k+1/2}(\alpha^{-2}) + B_{k-1/2}(\alpha^{-2})}{2B_{1/2}(\alpha^{-2})}\right],$$

where $B_\nu(\cdot)$ is defined in Example (1.7) and denotes the modified Bessel function of the third kind and order ν. The expected value, variance, skewness, and kurtosis are, respectively, $\mathbb{E}(T) = \eta(1 + \alpha^2/2)$, $\mathbb{VAR}(T) = (\alpha\eta)^2(1 + 5\alpha^2/4)$, $\mu_3 = 16\alpha^2(11\alpha^2 + 6)/(5\alpha^2 + 4)^3$, and $\mu_4 = 3 + 6\alpha^2(93\alpha^2 + 41)/(5\alpha^2 + 4)^2$. As noted before, if $T \sim \mathrm{BS}(\alpha, \eta)$, then $T^{-1} \sim \mathrm{BS}(\alpha, \eta^{-1})$. It then follows that $\mathbb{E}(T^{-1}) = \eta^{-1}(1 + \alpha^2/2)$ and $\mathbb{VAR}(T^{-1}) = \alpha^2\eta^{-2}(1 + 5\alpha^2/4)$.

The BS hazard rate function is given by

$$r(t) = \frac{\kappa(\alpha, \eta) t^{-3/2}(t + \eta)\exp[-\tau(t/\eta)/(2\alpha^2)]}{1 - \Phi(v_t)}, \quad t > 0.$$

Fig. 1.1 displays some plots of the BS density and hazard rate functions for selected values of α with $\eta = 1$. Notice that the BS density function is positively skewed and the asymmetry of the distribution decreases with α. As α decreases, the distribution becomes more symmetric around η, the median.

Little work has been published on the analysis of censored data for the BS distribution although censoring is common in reliability and survival studies. The readers are referred to Lemonte and Ferrari [8] and references therein. A particularly useful censoring mechanism is the type II (right) censoring, which occurs when n items are placed on test and the experiment is terminated when the first m $(m < n)$ items fail. Since life testing is often time consuming and expensive, type II censoring may be used to reduce testing time and costs. It is a convenient censoring scheme in which the number of observed failure times is known in advance. Also, it is the basis for more sophisticated censoring schemes such as the progressively type II censoring and the hybrid censoring scheme, which is a mixture of type I and type II censoring plans and provides a fixed time length for the experiment. In the following we focus on testing inference for the BS distribution parameters under type II censored samples.

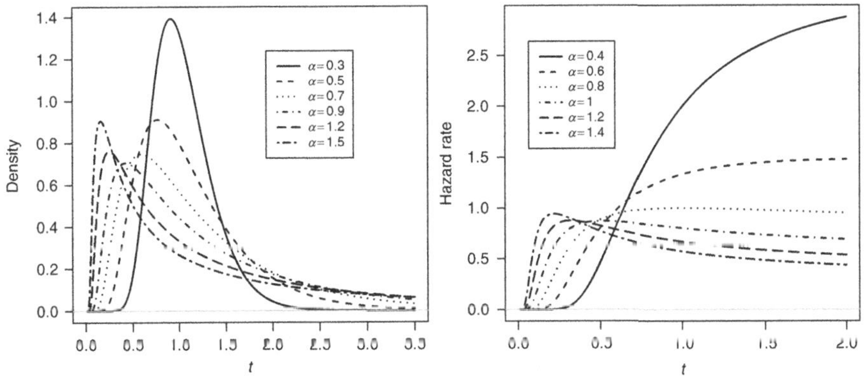

Fig. 1.1 Plots of the density and hazard rate functions of the BS distribution: $\eta = 1$.

1.5.1 Inference Under Type II Censored Samples

Let $t_1, \ldots, t_m$ be an ordered type II right censored sample obtained from n units placed on a life-testing experiment wherein each unit has its lifetime following the BS distribution, with the largest $(n - m)$ lifetimes having been censored. Let $\boldsymbol{\theta} = (\alpha, \eta)^\top$ be the parameter vector. For notational convenience, let

$$H(z) = \frac{\phi(z)}{1 - \Phi(z)}, \quad h(z) = \frac{\tau(z^{1/2})}{\alpha}, \quad t_l^* = \frac{t_l}{\eta},$$

where $\phi(\cdot)$ is the standard normal probability density function.

The likelihood function can be written as

$$Lik(\boldsymbol{\theta}) = \frac{n![1 - \Phi(v_m)]^{n-m}\kappa(\alpha, \eta)^m}{m!(n-m)!} \prod_{l=1}^{m} t_l^{-3/2}(t_l + \eta) \exp\left(-\frac{\tau(t_l^*)}{2\alpha^2}\right),$$

where $v_m = \rho(t_m^*)/\alpha$. Thus, the log-likelihood function, except for a constant term, is given by

$$\ell(\boldsymbol{\theta}) = \ell(\alpha, \eta) = m \log \kappa(\alpha, \eta) + \sum_{l=1}^{m} \log(t_l + \eta) - \frac{1}{2\alpha^2} \sum_{l=1}^{m} \tau(t_l^*)$$
$$+ (n - m) \log[1 - \Phi(v_m)].$$

By taking partial derivatives of the log-likelihood function with respect to α and η we obtain the components of the score vector, $\boldsymbol{U}(\boldsymbol{\theta}) = (U_\alpha, U_\eta)^\top$:

$$U_\alpha = U_\alpha(\alpha, \eta) = -\frac{m}{\alpha}\left(1 + \frac{2}{\alpha^2}\right) + \frac{1}{\alpha^3} \sum_{l=1}^{m} \tau(t_l^*) + \frac{(n-m)v_m H(v_m)}{\alpha},$$

$$U_\eta = U_\eta(\alpha, \eta) = -\frac{m}{2\eta} + \sum_{l=1}^{m} \frac{1}{t_l + \eta} + \frac{1}{2\alpha^2\eta} \sum_{l=1}^{m} \rho(t_l^{*2})$$
$$+ \frac{(n-m)h(t_m^*)H(v_m)}{2\eta}.$$

Setting U_α and U_η equal to zero yields the MLE $\hat{\boldsymbol{\theta}} = (\hat{\alpha}, \hat{\eta})^\top$ of $\boldsymbol{\theta} = (\alpha, \eta)^\top$. These equations cannot be solved analytically and statistical software can be used to solve them numerically via iterative methods.

We now consider hypothesis testing on the parameters α and η based on the gradient statistic. The interest lies in testing the null hypotheses

$$\mathcal{H}_{00} : \alpha = \alpha_0, \quad \mathcal{H}_{01} : \eta = \eta_0,$$

which are tested against $\mathcal{H}_{a0} : \alpha \neq \alpha_0$ and $\mathcal{H}_{a1} : \eta \neq \eta_0$, respectively. Here α_0 and η_0 are positive known scalars. For testing $\mathcal{H}_{00}$, the gradient statistic is given by

$$S_{\mathrm{T}(\alpha)} = U_\alpha(\alpha_0, \tilde{\eta})(\hat{\alpha} - \alpha_0),$$

where $\tilde{\eta}$ is the restricted MLE of η obtained from the maximization of $\ell(\boldsymbol{\theta})$ under $\mathcal{H}_{00}$. For testing $\mathcal{H}_{01}$, we have

$$S_{\mathrm{T}(\eta)} = U_\eta(\tilde{\alpha}, \eta_0)(\hat{\eta} - \eta_0).$$

Here, $\tilde{\alpha}$ is the restricted MLE of α obtained from the maximization of $\ell(\boldsymbol{\theta})$ under $\mathcal{H}_{01}$. The limiting distribution of these statistics is χ_1^2 under the respective null hypothesis.

A bias-corrected estimator for the shape parameter α under type II censoring was derived in Ng et al. [10] and is given by

$$\bar{\alpha} = \hat{\alpha}\left\{1 - \frac{1}{n}\left[1 + 2.5\left(1 - \frac{m}{n}\right)\right]\right\}^{-1}.$$

The authors showed through Monte Carlo simulations that $\bar{\alpha}$ is less biased than the original MLE, $\hat{\alpha}$, of α. Therefore, based on this estimator and on Theorem 1.3, we can define the adjusted gradient statistic $S^*_{\mathrm{T}(\alpha)} = U_\alpha(\alpha_0, \tilde{\eta})(\bar{\alpha} - \alpha_0)$ for testing the null hypothesis $\mathcal{H}_{00} : \alpha = \alpha_0$. The limiting distribution of $S^*_{\mathrm{T}(\alpha)}$ is also χ_1^2 under $\mathcal{H}_{00}$. It is expected that the gradient test that uses the statistic $S^*_{\mathrm{T}(\alpha)}$ have better size performance than the gradient test based on the original gradient statistic $S_{\mathrm{T}(\alpha)}$. A disadvantage of the adjusted gradient statistic $S^*_{\mathrm{T}(\alpha)}$ is that it can take on negative values. However, in order to avoid negative values we shall redefine $S^*_{\mathrm{T}(\alpha)}$ as

$$S^*_{\mathrm{T}(\alpha)} = \max\{0, U_\alpha(\alpha_0, \tilde{\eta})(\bar{\alpha} - \alpha_0)\}.$$

An adjusted gradient statistic for testing the null hypothesis $\mathcal{H}_{01} : \eta = \eta_0$ is not considered here because a bias-corrected estimator for η under type II censoring is not available. In the next section, we shall present an extensive Monte Carlo simulation study in order to evaluate the performance of the gradient tests presented in this section.

1.5.2 Numerical Results

In this section, we shall present the results of an extensive Monte Carlo simulation study in which we evaluate the finite sample performance of the gradient test for testing hypotheses on the parameters of the BS distribution under type II right censored samples. We also consider for the sake of comparison the LR statistic in the simulations. The LR statistics for testing $\mathcal{H}_{00} : \alpha = \alpha_0$ and $\mathcal{H}_{01} : \eta = \eta_0$ are given by

$$S_{\mathrm{LR}(\alpha)} = 2\left[\ell(\hat{\alpha}, \hat{\eta}) - \ell(\alpha_0, \tilde{\eta})\right], \quad S_{\mathrm{LR}(\eta)} = 2\left[\ell(\hat{\alpha}, \hat{\eta}) - \ell(\tilde{\alpha}, \eta_0)\right],$$

respectively. The limiting distribution of these statistics is χ_1^2 under the respective null hypothesis.

We set the degree of censoring (d.o.c.) at 0%, 10%, 30%, and 50%, the sample size at $n = 20$ and 40, and the shape parameter at $\alpha = 0.1, 0.5$, and 1.0. Without loss of generality, the scale parameter η was kept fixed at 1.0. The nominal levels of the tests were $\gamma = 10\%$ and 5%. The number of Monte Carlo replications was 10,000. All Monte Carlo simulation experiments were performed using the Ox matrix programming language, which is freely distributed for academic purposes and available at http://www.doornik.com.

Tables 1.2 and 1.3 present the null rejection rates for $\mathcal{H}_{00} : \alpha = \alpha_0$ of the tests which are based on the statistics $S_{\mathrm{LR}(\alpha)}$, $S_{\mathrm{T}(\alpha)}$ and $S^*_{\mathrm{T}(\alpha)}$ for $n = 20$ and $n = 40$, respectively, that is, the percentage of times that the corresponding statistics exceed the 10% and 5% upper points of the reference χ^2 distribution. Entries are percentages. The main findings are as follows. First, for complete data without censoring (d.o.c. = 0%), the tests based on $S_{\mathrm{T}(\alpha)}$ and $S^*_{\mathrm{T}(\alpha)}$ are less size distorted than the test that uses

Table 1.2 Null Rejection Rates (%) for $\mathcal{H}_{00} : \alpha = \alpha_0$ With $n = 20$

d.o.c.		$S_{\mathrm{LR}(\alpha)}$		$S_{\mathrm{T}(\alpha)}$		$S^*_{\mathrm{T}(\alpha)}$	
(%)	α_0	10%	5%	10%	5%	10%	5%
0	0.1	11.53	6.07	9.70	4.25	9.83	4.52
	0.5	11.48	6.15	9.65	4.21	9.81	4.45
	1.0	11.73	6.25	9.62	4.26	9.78	4.46
10	0.1	11.85	6.12	9.57	4.29	9.84	4.90
	0.5	11.85	6.15	9.59	4.29	9.81	4.85
	1.0	12.09	6.19	9.58	4.35	9.77	4.73
30	0.1	12.98	6.98	9.76	4.30	10.30	5.12
	0.5	13.04	7.06	9.82	4.28	10.14	5.04
	1.0	13.08	7.26	9.76	4.19	9.76	4.91
50	0.1	14.35	7.99	9.30	3.95	9.78	4.91
	0.5	14.18	7.92	9.20	3.89	9.65	4.85
	1.0	14.60	8.05	8.67	3.58	9.23	4.59

Table 1.3 Null Rejection Rates (%) for $\mathcal{H}_{00} : \alpha = \alpha_0$ With $n = 40$

d.o.c.		$S_{LR(\alpha)}$		$S_{T(\alpha)}$		$S^*_{T(\alpha)}$	
(%)	α_0	10%	5%	10%	5%	10%	5%
0	0.1	10.76	5.23	9.67	4.82	9.81	5.16
	0.5	10.75	5.24	9.71	4.80	9.85	5.12
	1.0	10.84	5.36	9.67	4.81	9.79	5.10
10	0.1	10.79	5.70	9.63	4.61	9.87	4.79
	0.5	10.78	5.67	9.63	4.56	9.80	4.73
	1.0	10.90	5.69	9.67	4.51	9.89	4.73
30	0.1	11.10	5.55	9.40	4.43	9.72	4.64
	0.5	10.98	5.50	9.38	4.30	9.66	4.73
	1.0	10.90	5.59	9.24	4.24	9.27	4.60
50	0.1	11.42	6.02	9.38	4.57	9.61	5.10
	0.5	11.39	6.12	9.33	4.55	9.57	5.03
	1.0	11.51	6.23	9.12	4.33	9.22	4.91

$S_{LR(\alpha)}$, in fact, they produce null rejection rates that are very close to the nominal levels in all cases considered. In addition, the adjusted gradient test is slightly superior than the original gradient test. For example, for $n = 20$, $\alpha_0 = 0.5$, $\gamma = 10\%$, and d.o.c. $= 0\%$, the null rejection rates are 11.48% ($S_{LR(\alpha)}$), 9.65% ($S_{T(\alpha)}$), and 9.81% ($S^*_{T(\alpha)}$). Second, the size distortion of all tests increases with the d.o.c., the LR test and the original gradient test displaying null rejection rates that are, respectively, greater and smaller than the nominal levels. The adjusted gradient test is the less size distorted in most of the cases. For instance, for $n = 20$, $\alpha_0 = 1.0$, and $\gamma = 5\%$, the null rejection rates are 6.19% ($S_{LR(\alpha)}$), 4.35% ($S_{T(\alpha)}$), and 4.73% ($S^*_{T(\alpha)}$) for d.o.c. $= 10\%$, and 7.26% ($S_{LR(\alpha)}$), 4.19% ($S_{T(\alpha)}$), and 4.91% ($S^*_{T(\alpha)}$) for d.o.c. $= 30\%$. Note that all tests become less size distorted as the sample size increases, as expected.

The null rejection rates of the tests that employ $S_{LR(\eta)}$ and $S_{T(\eta)}$ as test statistics for testing the null hypothesis $\mathcal{H}_{01} : \eta = 1$ are presented in Tables 1.4 and 1.5 for $n = 20$ and $n = 40$, respectively. For complete data without censoring (d.o.c. $= 0\%$), the gradient test is less size distorted than the LR in all cases. For example, for $n = 20$, $\alpha = 0.5$, and $\gamma = 10\%$, the null rejection rates are 11.98% ($S_{LR(\eta)}$) and 10.53% ($S_{T(\eta)}$). Additionally, the size distortion of the tests increases with the degree of censoring and, similarly to what occurs for testing on the parameter α, the LR test presented

Table 1.4 Null Rejection Rates (%) for $\mathcal{H}_{01} : \eta = 1$ With $n = 20$

d.o.c.		$S_{LR(\eta)}$		$S_{T(\eta)}$	
(%)	α	10%	5%	10%	5%
0	0.1	12.06	6.43	10.74	4.98
	0.5	11.98	6.52	10.53	4.94
	1.0	11.74	6.74	10.21	4.93
10	0.1	12.00	6.44	10.69	4.70
	0.5	11.96	6.40	10.22	4.63
	1.0	11.87	6.29	9.78	4.60
30	0.1	12.87	6.95	10.53	4.66
	0.5	12.99	6.99	9.98	4.39
	1.0	13.12	6.91	9.41	4.39
50	0.1	13.49	7.85	9.90	4.22
	0.5	13.57	7.75	9.47	4.24
	1.0	13.58	7.69	8.19	4.23

Table 1.5 Null Rejection Rates (%) for $\mathcal{H}_{01} : \eta = 1$ With $n = 40$

d.o.c.		$S_{LR(\eta)}$		$S_{T(\eta)}$	
(%)	α	10%	5%	10%	5%
0	0.1	11.00	5.43	10.38	4.92
	0.5	11.00	5.50	10.40	4.95
	1.0	11.23	5.81	10.33	5.08
10	0.1	10.82	5.59	10.17	4.76
	0.5	10.82	5.45	10.18	4.81
	1.0	11.09	5.64	10.05	4.80
30	0.1	11.17	5.60	9.97	4.55
	0.5	11.11	5.63	9.81	4.57
	1.0	11.07	5.69	9.36	4.45
50	0.1	11.70	5.98	10.08	4.49
	0.5	11.71	5.97	9.70	4.41
	1.0	11.74	6.03	9.02	4.33

a liberal behavior, while the gradient test was conservative in the majority of the cases. For instance, when $n = 20$, $\alpha = 0.5$, and $\gamma = 10\%$, the null rejection rates are 11.96% ($S_{LR(\eta)}$) and 10.22% ($S_{T(\eta)}$) for d.o.c. = 10%, and 13.57% ($S_{LR(\eta)}$) and 9.47% ($S_{T(\eta)}$) for d.o.c. = 50%. It should be noticed that the gradient test is less size distorted than the LR test in all cases considered. Again, the tests become less size distorted as the sample size increases.

Table 1.6 contains the nonnull rejection rates (powers) of the tests. We set $\alpha = 0.5$, $\eta = 1$, $n = 80, 120$, and 150, and the degree of censoring at 0%, 20%, and 40%. The rejection rates were obtained under the alternative hypotheses $\mathcal{H}_{a0} : \alpha = \delta_1$ and $\mathcal{H}_{a1} : \eta = \delta_2$, for different values of δ_1 and δ_2. The test that uses $S_{\mathrm{LR}(\alpha)}$ presented smaller powers for testing hypotheses on the parameter α than the tests which are based on the statistics $S_{\mathrm{T}(\alpha)}$ and $S^*_{\mathrm{T}(\alpha)}$. For example, when $n = 80$, d.o.c. $= 20\%$, and $\delta_1 = 0.58$, the nonnull rejection rates are 44.89% ($S_{\mathrm{LR}(\alpha)}$), 48.51% ($S_{\mathrm{T}(\alpha)}$), and 52.02% ($S^*_{\mathrm{T}(\alpha)}$). Additionally, the tests which use $S_{\mathrm{LR}(\eta)}$ and $S_{\mathrm{T}(\eta)}$ for testing hypothesis on the parameter η have similar powers, the gradient test being slightly more powerful than the LR test. For instance, when $n = 150$, d.o.c. $= 0\%$, and $\delta_2 = 1.10$, the nonnull rejection rates are 77.31% ($S_{\mathrm{LR}(\eta)}$) and 77.87% ($S_{\mathrm{T}(\eta)}$). Note that the powers of all tests decrease as the degree of censoring increases. We also note that the powers of the tests increase with n and also with δ_1 and δ_2, as expected.

A natural question at this point is why the Wald and Rao score tests were not included in the previous simulation experiments. Recall that the Wald and score statistics involve the (expected) Fisher information matrix, which cannot be obtained analytically for the BS distribution under type II censoring. A common practice in such a case is to use the observed information in the place of the expected information. We followed this approach and ran various simulation experiments including the Wald and score tests. The general conclusion is that these tests cannot be recommended for the following reasons. First, the Wald and score tests were markedly oversized. For example, for $n = 20$, $\gamma = 10\%$, and $\alpha = 0.75$, the null rejection rates of the Wald and score tests for testing the null hypothesis $\mathcal{H}_{01} : \eta = 1$ are, respectively, 13.79% and 15.61% for d.o.c. $= 10\%$, 16.73% and 15.44% for d.o.c. $= 30\%$, and 19.02% and 13.70% for d.o.c. $= 50\%$. Second, in the simulations the inverse of the observed information matrix frequently produced negative standard errors in censored samples.

In summary, the best performing test for α is the adjusted gradient test, that is, the one that uses the bias-corrected estimator of α. As far as hypothesis testing on η is concerned, the gradient test performs better than the LR test and therefore should be preferred.

Table 1.6 Nonnull Rejection Rates (%): $\alpha = 0.5$, $\eta = 1$, and $\gamma = 10\%$

	d.o.c.	$\mathcal{H}_{a0} : \alpha = \delta_1$				$\mathcal{H}_{a1} : \eta = \delta_2$		
n	(%)	δ_1	$S_{LR(\alpha)}$	$S_{T(\alpha)}$	$S^*_{T(\alpha)}$	δ_2	$S_{LR(\eta)}$	$S_{T(\eta)}$
80	0	0.50	10.71	10.11	10.25	1.00	10.53	10.20
		0.54	23.90	25.84	27.97	1.04	19.06	19.50
		0.58	56.16	58.99	62.18	1.10	54.60	55.66
	20	0.50	10.65	9.55	9.68	1.00	10.55	10.10
		0.54	19.34	20.93	23.62	1.04	18.67	19.27
		0.58	44.89	48.51	52.02	1.10	52.70	54.15
	40	0.50	10.93	9.92	9.72	1.00	10.30	9.51
		0.54	15.74	17.20	19.83	1.04	17.34	18.47
		0.58	33.78	37.30	40.92	1.10	47.69	50.16
120	0	0.50	10.53	10.17	10.18	1.00	10.62	10.37
		0.54	30.34	32.59	34.90	1.04	22.74	23.25
		0.58	72.73	74.76	76.52	1.10	69.69	70.35
	20	0.50	10.64	10.02	10.14	1.00	10.19	9.80
		0.54	24.58	26.43	28.95	1.04	22.78	23.46
		0.58	60.06	62.86	65.65	1.10	68.10	69.23
	40	0.50	10.65	9.90	9.81	1.00	10.21	9.65
		0.54	19.02	21.07	23.71	1.04	20.63	22.17
		0.58	45.24	48.49	52.05	1.10	60.84	62.71
150	0	0.50	9.98	9.72	9.76	1.00	10.15	10.00
		0.54	35.53	38.04	40.24	1.04	26.34	26.97
		0.58	80.99	82.36	83.63	1.10	77.31	77.87
	20	0.50	10.13	9.94	9.99	1.00	10.39	10.15
		0.54	28.30	30.38	33.01	1.04	25.52	26.39
		0.58	68.24	70.56	73.17	1.10	75.83	76.92
	40	0.50	10.51	10.06	9.85	1.00	10.33	10.02
		0.54	21.85	23.99	26.53	1.04	22.49	23.86
		0.58	53.25	56.23	59.28	1.10	68.41	70.32

1.5.3 Empirical Applications

In this section, we consider the gradient test in two real data sets. (We also consider the LR test.) The first data set corresponds to the lifetime (in hours) of 10 sustainers of a certain type. The observations are 152.7, 172.0, 172.5, 173.3, 193.0, 204.7, 216.5, 234.9, 262.6, and 422.6. Suppose we wish to test $\mathcal{H}_{00} : \alpha = 0.21$ against $\mathcal{H}_{a0} : \alpha \neq 0.21$, and $\mathcal{H}_{01} : \eta = 180$ against $\mathcal{H}_{a1} : \eta \neq 180$. Under different censoring proportions, the observed values of the different test statistics and the corresponding p-values are given in Table 1.7.

Table 1.7 Test Statistics (*p*-Values Between Parentheses)

	Inference on α		
m	$S_{LR(\alpha)}$	$S_{T(\alpha)}$	$S^{*}_{T(\alpha)}$
10	2.1646 (0.1412)	2.7944 (0.0946)	4.0043 (0.0454)
9	0.0770 (0.7814)	0.0728 (0.7873)	0.0000 (1.0000)
8	0.3307 (0.5653)	0.2911 (0.5895)	0.0000 (1.0000)
7	0.6732 (0.4119)	0.5514 (0.4578)	0.1472 (0.7013)
6	0.8471 (0.3574)	0.6620 (0.4159)	0.2234 (0.6365)

	Inference on η	
m	$S_{LR(\eta)}$	$S_{T(\eta)}$
10	2.9417 (0.0863)	2.7580 (0.0968)
9	3.2449 (0.0716)	2.9248 (0.0872)
8	3.1616 (0.0754)	2.8499 (0.0914)
7	2.9510 (0.0858)	2.7036 (0.1001)
6	2.6797 (0.1016)	2.5463 (0.1106)

Note that for complete data (without censoring) the LR test does not reject the null hypothesis $\mathcal{H}_{00}$ at any usual significance level, whereas the original and adjusted gradient tests reject the null hypothesis $\mathcal{H}_{00}$ at the 10% and 5% significance levels, respectively. Except for the case where $m = 10$ (no censoring), the LR and gradient tests lead to the same conclusion. For testing the null hypothesis $\mathcal{H}_{01}$, unless $m = 7$, the same decision is reached by the LR and gradient tests, that is, the null hypothesis is rejected at the 10% significance level for $m = 8, 9$, and 10, and is not rejected at any usual significance level for $m = 6$.

The second data set corresponds to the survival times (in days) of the first 7 of a sample of 10 mice after being inoculated with a culture of tuberculosis. The observations are 41, 44, 46, 54, 55, 58, and 60. Suppose we are interested in testing $\mathcal{H}_{00} : \alpha = 0.13$ against $\mathcal{H}_{a0} : \alpha \neq 0.13$, and $\mathcal{H}_{01} : \eta = 52$ against $\mathcal{H}_{a1} : \eta \neq 52$. We have $S_{LR(\alpha)} = 1.8304$ (p-value = 0.1761), $S_{T(\alpha)} = 2.4497$ (p-value = 0.1175), $S^{*}_{T(\alpha)} = 4.2366$ (p-value = 0.0396), $S_{LR(\eta)} = 1.0099$ (p-value = 0.3149), and $S_{T(\eta)} = 1.0235$ (p-value = 0.3117). Note that the LR and original gradient tests do not reject the null hypothesis $\mathcal{H}_{00}$ at any usual significance level, whereas the adjusted gradient test rejects the null hypothesis $\mathcal{H}_{00}$ at the 5% level. Additionally, the null hypothesis $\mathcal{H}_{01}$ is not rejected at any usual significance level based on the statistics $S_{LR(\eta)}$ and $S_{T(\eta)}$.

1.6 CENSORED EXPONENTIAL REGRESSION MODEL

The statistical analysis of lifetime data has become a topic of considerable interest to statisticians and workers in areas such as engineering, medicine, and social sciences. The field has expanded rapidly in the last two decades and publications on the subject can be found in the literature of several areas. In survival analysis, the interest focuses on a nonnegative random variable, say t, which is the time until a certain event of interest happens. In medical applications, this event can be the death of a patient suffering from a certain disease, and in a reliability analysis, this would typically be the time to failure of a machine or part of a machine. One refers to t as a lifetime, a survival time, or a failure time.

Among the parametric models, the exponential distribution is important in applications. One of the reasons for its importance is that the exponential distribution has constant failure rate. Additionally, this model was the first lifetime model for which statistical methods were extensively developed in the life testing literature. We shall assume that t follows an exponential distribution with probability density function in the form $f(t;\theta) = \theta^{-1}\exp(-t/\theta)$, where $t > 0$ and $\theta > 0$. The class of exponential regression models (ExpRMs) is based on the distribution of $y = \log t$ instead of t. In this case, y has an extreme value distribution with density function $f(y;\mu) = \exp(y-\mu)\exp[-\exp(y-\mu)]$, where $y \in \mathbb{R}$ and $\mu = \log\theta \in \mathbb{R}$. This regression model is one of the most important parametric regression models for the analysis of lifetime data and it is employed in several applications.

Censoring is very common in lifetime data because of time limits and other restrictions on data collection. In a engineering life test experiment, for example, it is usually not feasible to continue experimentation until all items under study have failed. In a survival study, patients follow-up may be lost and also data analysis is usually done before all patients have reached the event of interest. The partial information contained in the censored observations is just a lower bound on the lifetime distribution. Reliability studies usually finish before all units have failed, even making use of accelerated tests. This is a special source of difficulty in the analysis of reliability data. Such data are said to be censored at right and they arise when some units are still running at the time of the data analysis, removed from test before they fail or because they failed from an extraneous cause. Some mechanisms of censoring are identified in the literature as, for example, types I and II censoring.

From now on, we assume that $t_1, \ldots, t_n$ are stochastically independent random variables representing the failure times of n individuals. Assume also that $c_1, \ldots, c_n$ are fixed values representing the censoring times. We then observe $(z_1, \delta_1), \ldots, (z_n, \delta_n)$, where $z_l = \min\{t_l, c_l\}$, $\delta_l = I(t_l \leq c_l)$ is the failure indicator, and $I(\cdot)$ is the indicator function. This form of censoring has been termed generalized type I censoring and it is common in clinical studies since patients enter the study at different times and the terminal point of the study is predetermined by the investigator. Usual type I censoring mechanism is a particular case of the generalized type I censoring when $c_1 = \cdots = c_n = c$. Type II censoring happens when the study continues until the failure of the first r individuals, where r is some predetermined integer ($r \leq n$).

1.6.1 The Regression Model

Let $t_1, \ldots, t_n$ be n independent observations. The ExpRM is defined by assuming that

$$f(t_l; \theta_l) = \theta_l^{-1} \exp(-t_l/\theta_l), \quad l = 1, \ldots, n, \tag{1.4}$$

where $t_l > 0$ and $\theta_l > 0$. The most useful functional form of θ_l is $\theta_l = \exp(\boldsymbol{x}_l^\top \boldsymbol{\beta})$, where $\boldsymbol{x}_l^\top = (x_{l1}, \ldots, x_{lp})$ contains the lth observation on p covariates, and $\boldsymbol{\beta} = (\beta_1, \ldots, \beta_p)^\top$ is a vector of unknown parameters to be estimated from the data. An advantage of the exponential form for θ_l is that the requirement $\theta_l > 0$ is automatically satisfied. If we let $y_l = \log t_l$, then the ExpRM can be written in the form of the so-called accelerated lifetime model

$$y_l = \mu_l + \varepsilon_l, \quad l = 1, \ldots, n, \tag{1.5}$$

where $\mu_l = \log \theta_l = \boldsymbol{x}_l^\top \boldsymbol{\beta}$, and ε_l has a standard extreme value distribution with probability density function $f(\varepsilon_l) = \exp(\varepsilon_l) \exp[-\exp(\varepsilon_l)]$. This is a location-scale regression model with error variable ε_l.

For a censored sample based on n observations, y_l represents the observed log-lifetime or log-censoring, that is, for generalized type I censoring we have $y_l = \log(\min\{t_l, c_l\})$ and for type II censoring y_l becomes $y_l = \log(\min\{t_l, t_{(r)}\})$, where $t_{(r)}$ is the rth order statistic. Let $\ell(\boldsymbol{\beta})$ be the log-likelihood function for any model defined by Eq. (1.5) (or Eq. (1.4)) in some parameter space given an observed data set $(y_1, \delta_1), \ldots, (y_n, \delta_n)$. The function $\ell(\boldsymbol{\beta})$ for the two types of censoring has the same form, and it is given by

$$\ell(\boldsymbol{\beta}) = \sum_{l=1}^{n} [\delta_l(y_l - \mu_l) - \exp(y_l - \mu_l)]. \tag{1.6}$$

The score function for $\boldsymbol{\beta}$ is $\boldsymbol{U}(\boldsymbol{\beta}) = \boldsymbol{X}^\top \boldsymbol{W}^{1/2}\boldsymbol{v}$, where $\boldsymbol{X} = (\boldsymbol{x}_1, \ldots, \boldsymbol{x}_n)^\top$ is the model matrix of full rank (rank$(\boldsymbol{X}) = p$), $\boldsymbol{W} = \boldsymbol{W}(\boldsymbol{\beta}) = \mathrm{diag}\{w_1, \ldots, w_n\}$ with $w_l = \mathbb{E}[\exp(y_l - \mu_l)]$, and $\boldsymbol{v} = \boldsymbol{v}(\boldsymbol{\beta}) = (v_1, \ldots, v_n)^\top$ is an n-vector with typical element $v_l = w_l^{-1/2}[\exp(y_l - \mu_l) - \delta_l]$. The quantity w_l depends on the mechanism of censoring. We have $w_l = 1 - \exp[-c_l \exp(-\mu_l)]$ for generalized type I censoring and $w_l = r/n$ for type II censoring, where in this case r is a fixed number of failures.

The MLE $\hat{\boldsymbol{\beta}}$ of $\boldsymbol{\beta}$ can be obtained by iteratively solving the equation

$$\boldsymbol{X}^\top \boldsymbol{W}^{(m)} \boldsymbol{X} \boldsymbol{\beta}^{(m+1)} = \boldsymbol{X}^\top \boldsymbol{W}^{(m)} \boldsymbol{\zeta}^{(m)}, \quad m = 0, 1, \ldots, \tag{1.7}$$

where $\boldsymbol{\zeta}^{(m)} = \boldsymbol{X}\boldsymbol{\beta}^{(m)} + (\boldsymbol{W}^{(m)})^{-1/2}\boldsymbol{v}^{(m)}$ is an adjusted dependent variable. The procedure may be initialized by taking $\boldsymbol{\beta}^{(0)} = (\boldsymbol{X}^\top\boldsymbol{X})^{-1}\boldsymbol{X}^\top\boldsymbol{y}$, the least squares estimate of $\boldsymbol{\beta}$ without censoring, where $\boldsymbol{y} = (y_1, \ldots, y_n)^\top$. The cycles through the scheme (1.7) consist of iterative reweighted least squares steps and the iterations go on until convergence is achieved. Eq. (1.7) shows that the calculation of the MLE $\hat{\boldsymbol{\beta}}$ can be carried out using any software with a matrix algebra library.

Let $\mathcal{H}_0 : \boldsymbol{\beta}_1 = \boldsymbol{\beta}_{10}$ be the null hypothesis of interest, which will be tested against the alternative hypothesis $\mathcal{H}_a : \boldsymbol{\beta}_1 \neq \boldsymbol{\beta}_{10}$, where $\boldsymbol{\beta}$ is partitioned as $\boldsymbol{\beta} = (\boldsymbol{\beta}_1^\top, \boldsymbol{\beta}_2^\top)^\top$, $\boldsymbol{\beta}_1 = (\beta_1, \ldots, \beta_q)^\top$, and $\boldsymbol{\beta}_2 = (\beta_{q+1}, \ldots, \beta_p)^\top$. Here, $\boldsymbol{\beta}_{10}$ is a fixed column vector of dimension q, and $\boldsymbol{\beta}_2$ acts as a nuisance parameter vector. Let $\hat{\boldsymbol{\beta}} = (\hat{\boldsymbol{\beta}}_1^\top, \hat{\boldsymbol{\beta}}_2^\top)^\top$ and $\tilde{\boldsymbol{\beta}} = (\boldsymbol{\beta}_{10}^\top, \tilde{\boldsymbol{\beta}}_2^\top)^\top$ be the unrestricted and restricted MLEs of $\boldsymbol{\beta} = (\boldsymbol{\beta}_1^\top, \boldsymbol{\beta}_2^\top)^\top$, respectively. The gradient statistic to test the null hypothesis $\mathcal{H}_0 : \boldsymbol{\beta}_1 = \boldsymbol{\beta}_{10}$ can be expressed as

$$S_{\mathrm{T}} = \tilde{\boldsymbol{v}}^\top \tilde{\boldsymbol{W}}^{1/2} \boldsymbol{X}_1 (\hat{\boldsymbol{\beta}}_1 - \boldsymbol{\beta}_{10}), \tag{1.8}$$

where the matrix $\boldsymbol{X}$ is partitioned as $\boldsymbol{X} = [\boldsymbol{X}_1 \;\; \boldsymbol{X}_2]$, $\boldsymbol{X}_1$ and $\boldsymbol{X}_2$ being matrices of dimensions $n \times q$ and $n \times (p - q)$, respectively, $\tilde{\boldsymbol{v}} = \boldsymbol{v}(\tilde{\boldsymbol{\beta}})$ and $\tilde{\boldsymbol{W}} = \boldsymbol{W}(\tilde{\boldsymbol{\beta}})$. The limiting distribution of S_{T} is χ_q^2 under $\mathcal{H}_0$.

1.6.2 Finite-Sample Size Properties

In this section, we conduct Monte Carlo simulations in order to verify the performance of the gradient test in small- and moderate-sized samples for testing hypotheses in the class of ExpRMs. Both mechanisms of censoring are considered in the simulations. The simulations are performed for several combinations varying the sample sizes and the proportion of failures in the sample, P_{F} say. The simulation study is based on the regression model

$$y_l = \beta_1 x_{l1} + \beta_2 x_{l2} + \beta_3 x_{l3} + \beta_4 x_{l4} + \varepsilon_l,$$

where $x_{l1} = 1$, for $l = 1, \ldots, n$. The null hypothesis is $\mathcal{H}_0 : \beta_1 = \beta_2 = 0$, which is tested against a two-sided alternative hypothesis. The covariate values were selected as random draws from the $\mathcal{U}(0, 1)$ distribution and for fixed n those values were kept constant throughout the experiment. The number of Monte Carlo replications was 10,000 and the nominal levels of the tests were $\gamma = 10\%$ and 5%. The simulations were carried out using Ox matrix programming language (http://www.doornik.com). All parameters, except those fixed at the null hypothesis, were set equal to one.

For the sake of comparison we also consider in the simulations the LR statistic which assumes the form

$$S_{\mathrm{LR}} = 2 \sum_{l=1}^{n} \left[\delta_l (\tilde{\mu}_l - \hat{\mu}_l) - \mathrm{e}^{y_l} (\mathrm{e}^{-\hat{\mu}_l} - \mathrm{e}^{-\tilde{\mu}_l}) \right],$$

where tildes and hats indicate quantities available at the restricted and unrestricted MLEs, respectively. The limiting distribution of S_{LR} is χ^2_2 under the null hypothesis $\mathcal{H}_0 : \beta_1 = \beta_2 = 0$.

First, we present the results regarding the no censoring case. The null rejection rates of $\mathcal{H}_0 : \beta_1 = \beta_2 = 0$ for the two tests by considering different sample sizes are presented in Table 1.8. The LR and gradient tests are slightly liberal or conservative in most of the cases. However, they are very close to the corresponding nominal levels in all cases, which means that both tests present a very good performance in testing hypotheses in ExpRMs without censoring.

The simulation results for the types I and II censoring are presented in Tables 1.9 and 1.10, respectively. Note that the LR and gradient tests have similar null rejection rates in all cases. In general, when the proportion P_{F} of failures decreases, notice that the size distortion of all tests increases. The values in those tables show that the null rejection rates of the two tests approach the corresponding nominal levels as the sample size grows, as expected.

Table 1.8 Null Rejection Rates (%) for $\mathcal{H}_0 : \beta_1 = \beta_2 = 0$ Without Censoring

	$\gamma = 10\%$		$\gamma = 5\%$	
n	S_{LR}	S_T	S_{LR}	S_T
20	10.30	10.00	5.15	4.60
40	10.65	10.65	4.75	4.40
60	10.60	10.45	4.70	4.55
80	10.65	10.45	5.90	5.70
100	9.90	9.85	4.60	4.45
130	10.25	10.10	4.50	4.50
150	10.30	10.30	5.05	5.05

Table 1.9 Null Rejection Rates (%) for $\mathcal{H}_0 : \beta_1 = \beta_2 = 0$, Type I Censoring

		$\gamma = 10\%$		$\gamma = 5\%$	
P_F	n	S_{LR}	S_T	S_{LR}	S_T
90%	20	12.58	12.32	6.48	6.04
	30	11.66	11.44	6.42	6.32
	40	11.32	11.14	5.40	5.34
	50	11.28	11.18	5.74	5.76
	60	9.64	9.54	4.66	4.56
	70	10.48	10.36	5.48	5.38
	80	10.46	10.40	4.98	4.96
70%	20	15.38	15.46	8.58	8.48
	30	13.22	13.26	7.04	7.22
	40	12.12	12.42	6.52	6.66
	50	11.68	11.62	6.40	6.36
	60	11.54	11.58	5.96	5.94
	70	11.26	11.40	6.34	6.34
	80	10.74	10.68	5.54	5.60

In Table 1.11 we present the first four moments of S_{LR} and S_T and the corresponding moments of the limiting χ^2 distribution. We consider $P_F = 90\%$ in both cases of censoring. Note that the gradient statistic presents a good agreement between the true moments (obtained by simulation) and the moments of the limiting distribution.

Table 1.10 Null Rejection Rates (%) for $\mathcal{H}_0 : \beta_1 = \beta_2 = 0$, Type II Censoring

		$\gamma = 10\%$		$\gamma = 5\%$	
p_F	n	S_{LR}	S_T	S_{LR}	S_T
80%	20	13.76	13.58	7.40	7.14
	30	10.44	10.52	5.12	5.22
	40	11.74	11.72	6.22	6.12
	50	11.68	11.72	5.92	6.06
	60	10.88	10.88	5.68	5.68
	70	10.80	10.74	5.32	5.26
	80	10.64	10.80	5.72	5.78
60%	20	14.42	14.42	7.82	8.18
	30	11.98	12.54	6.08	6.48
	40	11.10	11.20	5.80	5.94
	50	11.68	11.70	6.34	6.36
	60	11.16	11.16	5.94	5.98
	70	11.32	11.34	5.84	5.88
	80	10.96	10.96	5.52	5.64

Table 1.11 Moments; $n = 50$

	No Censoring		Type I Censoring		Type II Censoring		
	S_{LR}	S_T	S_{LR}	S_T	S_{LR}	S_T	χ^2_2
Mean	1.97	1.95	2.14	2.13	2.00	1.99	2.0
Variance	3.65	3.57	4.45	4.45	3.88	3.80	4.0
Skewness	1.92	1.88	1.94	1.95	1.91	1.88	2.0
Kurtosis	8.48	8.22	8.57	8.76	7.94	7.81	9.0

1.6.3 An Empirical Application

We now consider the gradient test in a real data set reported in Nelson and Meeker [11] on insulating fluids in an accelerated test that was conducted in order to determine the relationship between time (in minutes) to breakdown (t) and voltage (x). In this experiment 76 units were tested at 7 stress levels, equally spaced between 26 kV and 38 kV. Nelson and Meeker [11] considered these data to find estimates of extreme value and Weibull distribution percentiles.

Nelson and Meeker [11] assumed a common regression coefficient for the seven groups and a common censoring time at $c = 200$ min. Following these authors we assume the regression model

$$y_l = \beta_1 + \beta_2 \log x_l + \varepsilon_l, \quad l = 1, \ldots, 76,$$

where $y_l = \log(\min\{t_l, c\})$. This regression model corresponds to an ExpRM with type I censoring involving a simple covariate. The MLEs of the parameters (standard errors between parentheses) are $\hat{\beta}_1 = 67.2400$ (5.6577) and $\hat{\beta}_2 = -18.2538$ (1.6069). We wish to test the significance of the voltage, that is, the null hypothesis of interest is $\mathcal{H}_0 : \beta_2 = 0$. We have $S_{\mathrm{T}} = 225.5901$ and hence based on this observed value we reject the null hypothesis under common critical values.

CHAPTER 2

The Local Power of the Gradient Test

2.1 PRELIMINARIES

Let $\ell(\boldsymbol{\theta})$ be the log-likelihood function, and $\boldsymbol{\theta} = (\boldsymbol{\theta}_1^\top, \boldsymbol{\theta}_2^\top)^\top$, where $\boldsymbol{\theta}_1$ and $\boldsymbol{\theta}_2$ are parameter vectors of dimensions q and $p-q$, respectively. Consider the problem of testing the composite null hypothesis $\mathcal{H}_0 : \boldsymbol{\theta}_1 = \boldsymbol{\theta}_{10}$ against the two-sided alternative hypothesis $\mathcal{H}_a : \boldsymbol{\theta}_1 \neq \boldsymbol{\theta}_{10}$, where $\boldsymbol{\theta}_{10}$ is a q-dimensional fixed vector, and $\boldsymbol{\theta}_2$ acts as a parameter vector. The gradient statistic for testing the null hypothesis $\mathcal{H}_0 : \boldsymbol{\theta}_1 = \boldsymbol{\theta}_{10}$ is defined in Eq. (1.3).

We introduce the following log-likelihood derivatives

$$y_r = n^{-1/2}\frac{\partial \ell(\boldsymbol{\theta})}{\partial \theta_r}, \quad y_{rs} = n^{-1}\frac{\partial^2 \ell(\boldsymbol{\theta})}{\partial \theta_r \partial \theta_s}, \quad y_{rst} = n^{-3/2}\frac{\partial^3 \ell(\boldsymbol{\theta})}{\partial \theta_r \partial \theta_s \partial \theta_t},$$

their arrays

$$\boldsymbol{y} = (y_1, \ldots, y_p)^\top, \quad \boldsymbol{Y} = ((y_{rs})), \quad \boldsymbol{Y}_{\ldots} = ((y_{rst})),$$

The Gradient Test. http://dx.doi.org/10.1016/B978-0-12-803596-2.00002-8

the corresponding cumulants

$$\kappa_{rs} = \mathbb{E}(y_{rs}), \quad \kappa_{r,s} = \mathbb{E}(y_r y_s), \quad \kappa_{rst} = n^{1/2}\mathbb{E}(y_{rst}),$$

$$\kappa_{r,st} = n^{1/2}\mathbb{E}(y_r y_{st}), \quad \kappa_{r,s,t} = n^{1/2}\mathbb{E}(y_r y_s y_t),$$

and their arrays

$$\boldsymbol{K} = \mathbb{E}(\boldsymbol{y}\boldsymbol{y}^\top) = ((\kappa_{r,s})), \quad \boldsymbol{K}_{...} = ((\kappa_{rst})),$$

$$\boldsymbol{K}_{.,..} = ((\kappa_{r,st})), \quad \boldsymbol{K}_{.,.,.} = ((\kappa_{r,s,t})),$$

where $r, s, t = 1, \ldots, p$.

We have that the κs are not functionally independent; for instance,

$$\kappa_{r,s} + \kappa_{rs} = 0, \quad \kappa_{rst} + \kappa_{rs,t} + \kappa_{rt,s} + \kappa_{st,r} = -\kappa_{r,s,t},$$

$$\kappa_{r,st} + \kappa_{rst} = \frac{\partial \kappa_{st}}{\partial \theta_r},$$

etc. Also, it is assumed that $\boldsymbol{Y}$ is non-singular and that $\boldsymbol{K}$ is positive definite with inverse $\boldsymbol{K}^{-1} = ((\kappa^{r,s}))$ say. We use the following summation notation:

$$\boldsymbol{K}_{...} \circ \boldsymbol{a} \circ \boldsymbol{b} \circ \boldsymbol{c} = \sum_{r,s,t=1}^{p} \kappa_{rst} a_r b_s c_t, \quad \boldsymbol{K}_{.,..} \circ \boldsymbol{M} \circ \boldsymbol{b} = \sum_{r,s,t=1}^{p} \kappa_{r,st} m_{rs} b_t,$$

where $\boldsymbol{M} = ((m_{rs}))$ is a $p \times p$ matrix, and $\boldsymbol{a} = (a_1, \ldots, a_p)^\top$, $\boldsymbol{b} = (b_1, \ldots, b_p)^\top$, and $\boldsymbol{c} = (c_1, \ldots, c_p)^\top$ are p-dimensional column vectors. The partition $\boldsymbol{\theta} = (\boldsymbol{\theta}_1^\top, \boldsymbol{\theta}_2^\top)^\top$ induces the corresponding partitions:

$$\boldsymbol{Y} = \begin{bmatrix} \boldsymbol{Y}_{11} & \boldsymbol{Y}_{12} \\ \boldsymbol{Y}_{21} & \boldsymbol{Y}_{22} \end{bmatrix}, \quad \boldsymbol{K} = \begin{bmatrix} \boldsymbol{K}_{11} & \boldsymbol{K}_{12} \\ \boldsymbol{K}_{21} & \boldsymbol{K}_{22} \end{bmatrix}, \quad \boldsymbol{K}^{-1} = \begin{bmatrix} \boldsymbol{K}^{11} & \boldsymbol{K}^{12} \\ \boldsymbol{K}^{21} & \boldsymbol{K}^{22} \end{bmatrix}.$$

Let $\mathcal{H}_{an} : \boldsymbol{\theta}_1 = \boldsymbol{\theta}_{10} + n^{-1/2}\boldsymbol{\epsilon}$ be the sequence of Pitman (local) alternative hypotheses converging to the null hypothesis at rate $n^{-1/2}$, where $\boldsymbol{\epsilon} = (\epsilon_1, \ldots, \epsilon_q)^\top$. We can express the asymptotic expansion of S_{T} for the composite hypothesis up to an error of order $O_p(n^{-1})$ as

$$\begin{aligned} S_{\mathrm{T}} = &-(\boldsymbol{Z}\boldsymbol{y} - \boldsymbol{\xi})^\top \boldsymbol{Y}(\boldsymbol{Z}\boldsymbol{y} - \boldsymbol{\xi}) \\ &- \frac{1}{2\sqrt{n}}\boldsymbol{K}_{...} \circ (\boldsymbol{Z}\boldsymbol{y} - \boldsymbol{\xi}) \circ \boldsymbol{Y}^{-1}\boldsymbol{y} \circ \boldsymbol{Y}^{-1}\boldsymbol{y} \\ &- \frac{1}{2\sqrt{n}}\boldsymbol{K}_{...} \circ (\boldsymbol{Z}\boldsymbol{y} - \boldsymbol{\xi}) \circ (\boldsymbol{Z}_0\boldsymbol{y} + \boldsymbol{\xi}) \circ (\boldsymbol{Z}_0\boldsymbol{y} + \boldsymbol{\xi}) \\ &+ O_p(n^{-1}), \end{aligned}$$

where $\boldsymbol{Z} = \boldsymbol{Y}^{-1} - \boldsymbol{Z}_0$,

$$\boldsymbol{Z}_0 = \begin{bmatrix} \mathbf{0}_{q,q} & \mathbf{0}_{q,\,p-q} \\ \mathbf{0}_{p-q,q} & \boldsymbol{Y}_{22}^{-1} \end{bmatrix}, \quad \boldsymbol{\xi} = \begin{bmatrix} \boldsymbol{I}_q \\ -\boldsymbol{Y}_{22}^{-1}\boldsymbol{Y}_{21} \end{bmatrix} \boldsymbol{\epsilon},$$

and $\boldsymbol{\epsilon} = \sqrt{n}(\boldsymbol{\theta}_1 - \boldsymbol{\theta}_{10})$. Also, $\mathbf{0}_{l,c}$ denotes an $l \times c$ matrix of zeros.

We can now use a multivariate Edgeworth type A series expansion of the joint density function of $\boldsymbol{y}$ and $\boldsymbol{Y}$ up to order $O(n^{-1/2})$, which has the form

$$f_1 = f_0 \left[1 + \frac{1}{6\sqrt{n}} \left(\boldsymbol{K}_{,,,} \circ \boldsymbol{K}^{-1}\boldsymbol{y} \circ \boldsymbol{K}^{-1}\boldsymbol{y} \circ \boldsymbol{K}^{-1}\boldsymbol{y} \right.\right.$$
$$\left.\left. -3\boldsymbol{K}_{,,,} \circ \boldsymbol{K}^{-1} \circ \boldsymbol{K}^{-1}\boldsymbol{y} - 6\boldsymbol{K}_{,,,} \circ \boldsymbol{K}^{-1}\boldsymbol{y} \circ \boldsymbol{D} \right) \right]$$
$$+ O(n^{-1}),$$

where

$$f_0 = (2\pi)^{-p/2} \det(\boldsymbol{K})^{-1/2} \exp\left(-\frac{1}{2}\boldsymbol{y}^\top \boldsymbol{K}^{-1}\boldsymbol{y} \right) \prod_{r,s=1}^{p} \delta(y_{rs} - \kappa_{rs}),$$

$\boldsymbol{D} = ((d_{bc}))$, $d_{bc} = \delta'(y_{bc} - \kappa_{bc})/\delta(y_{bc} - \kappa_{bc})$, $\delta(\cdot)$ is the Dirac delta function and $\det(\boldsymbol{K})$ denotes the determinant of the matrix $\boldsymbol{K}$, to obtain the moment generating function of S_{T}, $M(t)$ say.

From f_1 and the asymptotic expansion of S_{T} up to order $O_p(n^{-1/2})$, we arrive at

$$M(t) = (1-2t)^{-q/2} \exp\left(\frac{t}{1-2t} \boldsymbol{\epsilon}^\top \boldsymbol{K}_{11.2} \boldsymbol{\epsilon} \right)$$
$$\times \left[1 + \frac{1}{\sqrt{n}} (A_1 d + A_2 d^2 + A_3 d^3) \right] + O(n^{-1}),$$

where $d = 2t/(1-2t)$, $\boldsymbol{K}_{11.2} = \boldsymbol{K}_{11} - \boldsymbol{K}_{12}\boldsymbol{K}_{22}^{-1}\boldsymbol{K}_{21}$,

$$A_1 = \frac{1}{4} \sum_{r,s,t=1}^{p} \kappa_{rst}\kappa^{r,s}\epsilon_t^* + \frac{1}{4} \sum_{r,s,t=1}^{p} (\kappa_{rst} + 4\kappa_{r,st}) a_{rs}\epsilon_t^*$$
$$+ \frac{1}{4} \sum_{r,s,t=1}^{p} \kappa_{rst}\epsilon_r^*\epsilon_s^*\epsilon_t^*,$$

$$A_2 = \frac{1}{4} \sum_{r,s,t=1}^{p} \kappa_{rst}\kappa^{r,s}\epsilon_t^* - \frac{1}{4} \sum_{r,s,t=1}^{p} \kappa_{rst} a_{rs}\epsilon_t^*$$
$$- \frac{1}{2} \sum_{r,s,t=1}^{p} \kappa_{rst}\epsilon_r^*\epsilon_s^*\epsilon_t^*,$$

$$A_3 = \frac{1}{12}\sum_{r,s,t=1}^{p} \kappa_{rst}\epsilon_r^*\epsilon_s^*\epsilon_t^*, \quad \boldsymbol{\epsilon}^* = \begin{pmatrix}\epsilon_1^* \\ \vdots \\ \epsilon_p^*\end{pmatrix} = \begin{bmatrix}\boldsymbol{I}_q \\ -\boldsymbol{K}_{22}^{-1}\boldsymbol{K}_{21}\end{bmatrix}\boldsymbol{\epsilon},$$

$$\boldsymbol{A} = ((a_{rs}))_{r,s=1\ldots,p} = \begin{bmatrix}\boldsymbol{0}_{q,q} & \boldsymbol{0}_{q,p-q} \\ \boldsymbol{0}_{p-q,q} & \boldsymbol{K}_{22}^{-1}\end{bmatrix}.$$

The readers are referred to Lemonte and Ferrari [12] and references therein for more details.

Notice that when $n \to \infty$,

$$M(t) \to (1-2t)^{-q/2}\exp\left(\lambda\frac{2t}{1-2t}\right),$$

where $\lambda = (1/2)\boldsymbol{\epsilon}^\top\boldsymbol{K}_{11.2}\boldsymbol{\epsilon}$, and hence the limiting distribution of S_T is a noncentral χ^2 distribution with q degrees of freedom and noncentrality parameter λ ($\chi^2_{q,\lambda}$). Under $\mathcal{H}_0 : \boldsymbol{\theta}_1 = \boldsymbol{\theta}_{10}$, that is, when $\boldsymbol{\epsilon} = \boldsymbol{0}_q$,

$$M(t) = (1-2t)^{-q/2} + O(n^{-1}),$$

and, as expected, the gradient statistic S_T has a central χ^2 distribution with q degrees of freedom up to an error of order $O(n^{-1})$. Also, from $M(t)$ we may obtain the first three moments (mean, variance, and third central moment) of S_T up to order $O(n^{-1/2})$ as

$$\mathbb{E}(S_\text{T}) = q + \lambda + \frac{2}{\sqrt{n}}A_1,$$

$$\mathbb{VAR}(S_\text{T}) = 2(q+2\lambda) + \frac{8}{\sqrt{n}}(A_1 + A_2),$$

$$\mu_3(S_\text{T}) = 8(q+3\lambda) + \frac{6}{\sqrt{n}}(A_1 + 2A_2 + A_3).$$

2.2 NONNULL DISTRIBUTION UP TO ORDER $O(n^{-1/2})$

The moment generating function of S_T in a neighborhood of $\boldsymbol{\theta}_1 = \boldsymbol{\theta}_{10}$ can be expressed, after some algebra, as

$$M(t) = (1-2t)^{-q/2}\exp\left(\frac{t}{1-2t}\boldsymbol{\epsilon}^\top\boldsymbol{K}_{11.2}\boldsymbol{\epsilon}\right) \times \left[1 + \frac{1}{\sqrt{n}}\sum_{k=0}^{3} b_k(1-2t)^{-k}\right] + O(n^{-1}),$$

where

$$\begin{aligned}
b_1 &= -\frac{1}{4}\sum_{r,s,t=1}^{p} \kappa_{rst}\kappa^{r,s}\epsilon_t^* + \frac{1}{2}\sum_{r,s,t=1}^{p}(\kappa_{rst}+2\kappa_{r,st})\epsilon_r^*\epsilon_s^*\epsilon_t^* \\
&\quad + \frac{1}{4}\sum_{r,s,t=1}^{p}(4\kappa_{r,st}+3\kappa_{rst})a_{rs}\epsilon_t^* - \frac{1}{2}\sum_{r=1}^{q}\sum_{s,t=1}^{p}(\kappa_{rst}+\kappa_{r,st})\epsilon_r\epsilon_s^*\epsilon_t^*, \\
b_2 &= \frac{1}{4}\sum_{r,s,t=1}^{p}\kappa_{rst}(\kappa^{r,s}-a_{rs})\epsilon_t^* - \frac{1}{4}\sum_{r,s,t=1}^{p}(\kappa_{rst}+2\kappa_{r,st})\epsilon_r^*\epsilon_s^*\epsilon_t^*, \\
b_3 &= \frac{1}{12}\sum_{r,s,t=1}^{p}\kappa_{rst}\epsilon_r^*\epsilon_s^*\epsilon_t^*,
\end{aligned} \tag{2.1}$$

and $b_0 = -(b_1+b_2+b_3)$. Inverting $M(t)$, we arrive at the following theorem.

Theorem 2.1. *The asymptotic expansion of the cumulative distribution function of the gradient statistic for testing a composite hypothesis under a sequence of local alternatives converging to the null hypothesis at rate $n^{-1/2}$ is*

$$\Pr(S_{\mathrm{T}} \le x|\mathcal{H}_{\mathrm{an}}) = G_{q,\lambda}(x) + \frac{1}{\sqrt{n}}\sum_{k=0}^{3} b_k G_{q+2k,\lambda}(x) + O(n^{-1}), \tag{2.2}$$

where $G_{\nu,\lambda}(\cdot)$ is the cumulative distribution function of a noncentral χ^2 variate with ν degrees of freedom and noncentrality parameter λ. Here, $\lambda = (1/2)\boldsymbol{\epsilon}^\top \boldsymbol{K}_{11.2}\boldsymbol{\epsilon}$ and the b_ks are given in Eq. (2.1).

If $q = p$, the null hypothesis is simple, $\boldsymbol{\epsilon}^* = \boldsymbol{\epsilon}$ and $\boldsymbol{A} = \boldsymbol{0}_{p,p}$. Therefore, an immediate consequence of Theorem 2.1 is the following corollary.

Corollary 2.1. *The asymptotic expansion of the cumulative distribution function of the gradient statistic for testing a simple hypothesis under a sequence of local alternatives converging to the null hypothesis at rate $n^{-1/2}$ is given by*

$$\Pr(S_{\mathrm{T}} \le x|\mathcal{H}_{\mathrm{an}}) = G_{p,\lambda}(x) + \frac{1}{\sqrt{n}}\sum_{k=0}^{3} b_k G_{p+2k,\lambda}(x) + O(n^{-1}), \tag{2.3}$$

where $\lambda = (1/2)\boldsymbol{\epsilon}^\top \boldsymbol{K}\boldsymbol{\epsilon}$,

$$b_1 = -\frac{1}{4}\sum_{r,s,t=1}^{p} \kappa_{rst}\kappa^{r,s}\epsilon_t + \frac{1}{2}\sum_{r,s,t=1}^{p} \kappa_{r,st}\epsilon_r\epsilon_s\epsilon_t,$$

$$b_2 = \frac{1}{4}\sum_{r,s,t=1}^{p} \kappa_{rst}\kappa^{r,s}\epsilon_t - \frac{1}{4}\sum_{r,s,t=1}^{p} (\kappa_{rst} + 2\kappa_{r,st})\epsilon_r\epsilon_s\epsilon_t, \qquad (2.4)$$

$$b_3 = \frac{1}{12}\sum_{r,s,t=1}^{p} \kappa_{rst}\epsilon_r\epsilon_s\epsilon_t,$$

and $b_0 = -(b_1 + b_2 + b_3)$.

2.3 POWER COMPARISONS BETWEEN THE RIVAL TESTS

To first order, the gradient, LR, Wald, and score statistics have the same asymptotic distributional properties under either the null or local alternative hypotheses. Up to an error of order $O(n^{-1})$, the corresponding criteria have the same size but their powers differ in the $O(n^{-1/2})$ term. The power performance of the different tests may then be compared based on the expansions of their power functions ignoring terms of order less than $O(n^{-1/2})$. Harris and Peers [13] have presented a study of local power, up to order $O(n^{-1/2})$, for the LR, Wald, and score tests. They showed that none of the criteria is uniformly better than the others in terms of second order local power.

Let Π_i be the power function, up to order $O(n^{-1/2})$, of the test that uses the statistic S_i for $i = \text{LR, W, R, T}$. We can write the local powers as $\Pi_i = 1 - \Pr(S_i \le x|\mathcal{H}_{\text{an}}) = \Pr(S_i > x|\mathcal{H}_{\text{an}})$, where x is replaced by $\chi^2_q(\gamma)$ (ie, the upper $100(1-\gamma)\%$ quantile of the χ^2_q distribution for a given nominal level γ), and

$$\Pr(S_i \le x|\mathcal{H}_{\text{an}}) = G_{q,\lambda}(x) + \frac{1}{\sqrt{n}}\sum_{k=0}^{3} b_{ik}G_{q+2k,\lambda}(x) + O(n^{-1}).$$

The coefficients that define the local powers of the LR and Wald tests are given in Hayakawa [14], those corresponding to the score and gradient tests are given in Harris and Peers [13] and in Eq. (2.1), respectively. All of them are complicated functions of joint cumulants of log-likelihood derivatives but we can draw the following general conclusions:

- All four tests are locally biased in general.
- If $\kappa_{rst} = 0$ $(r, s, t = 1, \ldots, p)$, then the LR, Wald, and gradient tests have identical local power properties.
- If $\kappa_{rst} = 2\kappa_{r,s,t}$ $(r, s, t = 1, \ldots, p)$, then the score and gradient tests have identical local power properties.

Further classifications are possible for appropriate subspaces of the parameter space. Therefore, there is no uniform superiority of one test with respect to the others. Hence, the test that uses the gradient statistic, which is very simple to compute, can be an interesting alternative to the classic large-sample tests, namely the LR, Wald, and Rao score tests; that is, the gradient test is competitive with the other three tests since none is uniformly superior to the others in terms of second order local power.

2.4 NONNULL DISTRIBUTION UNDER ORTHOGONALITY

In this section, we examine the local power of the gradient test under the presence of a scalar parameter, ϕ say, that is orthogonal to the remaining parameters. We show that some of the coefficients that define the local power function of the gradient test remain unchanged regardless of whether ϕ is known or needs to be estimated, whereas the others can be written as the sum of two terms, the first of which being the corresponding term obtained as if ϕ were known, and the second, an additional term yielded by the fact that ϕ is unknown. The contribution of each set of parameters on the local power of the gradient test can then be examined. Various implications of this result are stated and discussed. We also present two examples for illustrative purposes.

Suppose $\boldsymbol{\theta} = (\boldsymbol{\beta}^\top, \phi)^\top$ is a $(p+1)$-vector of unknown parameters. Let $\boldsymbol{\beta} = (\boldsymbol{\beta}_1^\top, \boldsymbol{\beta}_2^\top)^\top$, where the dimensions of $\boldsymbol{\beta}_1$ and $\boldsymbol{\beta}_2$ are q and $p - q$, respectively, and ϕ is a scalar parameter. We focus on testing the composite null hypothesis $\mathcal{H}_0 : \boldsymbol{\beta}_1 = \boldsymbol{\beta}_{10}$ against the two-sided alternative hypothesis $\mathcal{H}_\text{a} : \boldsymbol{\beta}_1 \neq \boldsymbol{\beta}_{10}$, where $\boldsymbol{\beta}_{10}$ is a specified vector, and $\boldsymbol{\beta}_2$ and ϕ are nuisance parameters. We shall assume that $\boldsymbol{\beta} = (\boldsymbol{\beta}_1^\top, \boldsymbol{\beta}_2^\top)^\top$ is globally orthogonal to the parameter ϕ in the sense of Cox and Reid [15]:

$$\mathbb{E}\left(\frac{\partial^2 \ell(\boldsymbol{\theta})}{\partial \boldsymbol{\beta} \partial \phi}\right) = \mathbf{0}_p.$$

In other words, the Fisher information matrix for $\boldsymbol{\theta} = (\boldsymbol{\beta}^\top, \phi)^\top$ and its inverse are block-diagonal:

$$\boldsymbol{K}(\boldsymbol{\theta}) = \text{diag}\{\boldsymbol{K}_\beta, K_\phi\}, \quad \boldsymbol{K}(\boldsymbol{\theta})^{-1} = \text{diag}\{\boldsymbol{K}_\beta^{-1}, K_\phi^{-1}\},$$

where $\boldsymbol{K}_\beta$ is the Fisher information matrix for $\boldsymbol{\beta}$, and K_ϕ is the information relative to ϕ. There are numerous statistical models for which global orthogonality holds. From the partition of $\boldsymbol{\beta}$ we have that

$$\boldsymbol{K}_\beta = \begin{bmatrix} \boldsymbol{K}_{\beta 11} & \boldsymbol{K}_{\beta 12} \\ \boldsymbol{K}_{\beta 21} & \boldsymbol{K}_{\beta 22} \end{bmatrix}, \quad \boldsymbol{K}_\beta^{-1} = \begin{bmatrix} \boldsymbol{K}^{\beta 11} & \boldsymbol{K}^{\beta 12} \\ \boldsymbol{K}^{\beta 21} & \boldsymbol{K}^{\beta 22} \end{bmatrix}.$$

The gradient statistic for testing $\mathcal{H}_0 : \boldsymbol{\beta}_1 = \boldsymbol{\beta}_{10}$ takes the form

$$S_T = \boldsymbol{U}_{\beta_1}(\tilde{\boldsymbol{\theta}})^\top (\hat{\boldsymbol{\beta}}_1 - \boldsymbol{\beta}_{10}),$$

where $\hat{\boldsymbol{\theta}} = (\hat{\boldsymbol{\beta}}_1^\top, \hat{\boldsymbol{\beta}}_2^\top, \hat{\phi})^\top$ and $\tilde{\boldsymbol{\theta}} = (\boldsymbol{\beta}_{10}^\top, \tilde{\boldsymbol{\beta}}_2^\top, \tilde{\phi})^\top$ denote the unrestricted and restricted MLEs of $\boldsymbol{\theta} = (\boldsymbol{\beta}_1^\top, \boldsymbol{\beta}_2^\top, \phi)^\top$, respectively, and $\boldsymbol{U}_{\beta_1}(\boldsymbol{\theta}) = \partial \ell(\boldsymbol{\theta})/\partial \boldsymbol{\beta}_1$.

We shall assume the local alternative hypothesis $\mathcal{H}_{an} : \boldsymbol{\beta}_1 = \boldsymbol{\beta}_{10} + n^{-1/2}\boldsymbol{\epsilon}$, where $\boldsymbol{\epsilon} = (\epsilon_1, \ldots, \epsilon_q)^\top = \sqrt{n}(\boldsymbol{\beta}_1 - \boldsymbol{\beta}_{10})$. We define the quantities

$$\boldsymbol{\epsilon}^* = (\epsilon_1^*, \ldots, \epsilon_{p+1}^*)^\top = \begin{bmatrix} \boldsymbol{I}_q \\ -\boldsymbol{K}_{\beta 22}^{-1}\boldsymbol{K}_{\beta 21} \\ \boldsymbol{0}_{1,q} \end{bmatrix} \boldsymbol{\epsilon},$$

$$\boldsymbol{A} = ((a_{rs}))_{r,s=1,\ldots,p+1} = \begin{bmatrix} \boldsymbol{A}_\beta & \boldsymbol{0}_{p,1} \\ \boldsymbol{0}_{1,p} & K_\phi^{-1} \end{bmatrix},$$

$$\boldsymbol{M} = ((m_{rs}))_{r,s=1,\ldots,p+1} = \begin{bmatrix} \boldsymbol{M}_\beta & \boldsymbol{0}_{p,1} \\ \boldsymbol{0}_{1,p} & 0 \end{bmatrix},$$

$$\boldsymbol{M}_\beta = \boldsymbol{K}_\beta^{-1} - \boldsymbol{A}_\beta, \quad \boldsymbol{A}_\beta = \begin{bmatrix} \boldsymbol{0}_{q,q} & \boldsymbol{0}_{q,p-q} \\ \boldsymbol{0}_{p-q,q} & \boldsymbol{K}_{\beta 22}^{-1} \end{bmatrix}.$$

Note that whenever an index equals $p + 1$, it refers to ϕ, the last component of the parameter vector $\boldsymbol{\theta}$. To make the notation more intuitive, in many instances we will write ϕ instead of $p + 1$. We then have $m_{r\phi} = m_{\phi r} = m_{\phi\phi} = 0$, $a_{r\phi} = a_{\phi r} = 0$ $(r = 1, \ldots, p)$ and $a_{\phi\phi} = K_\phi^{-1}$.

By exploiting the orthogonality between $\boldsymbol{\beta}$ and ϕ, we arrive, after long and tedious algebraic manipulations, at the following general result.

Theorem 2.2. *Let $\boldsymbol{\theta} = (\boldsymbol{\beta}_1^\top, \boldsymbol{\beta}_2^\top, \phi)^\top$ be the parameter vector of dimension $p + 1$, where the dimensions of $\boldsymbol{\beta}_1$ and $\boldsymbol{\beta}_2$ are q and $p - q$,* *respectively, and ϕ is a scalar parameter. Assume that $\boldsymbol{\beta} = (\boldsymbol{\beta}_1^\top, \boldsymbol{\beta}_2^\top)^\top$ and ϕ are globally orthogonal. The nonnull asymptotic expansion of the cumulative distribution function of the gradient statistic for testing the null hypothesis $\mathcal{H}_0 : \boldsymbol{\beta}_1 = \boldsymbol{\beta}_{10}$ under a sequence of Pitman alternatives is given by*

$$\Pr(S_{\mathrm{T}} \leq x | \mathcal{H}_{\mathrm{an}}) = G_{q,\lambda}(x) + \frac{1}{\sqrt{n}} \sum_{k=0}^{3} b_k^* G_{q+2k,\lambda}(x) + O(n^{-1}),$$

where

$$b_1^* = b_1 + \xi, \quad b_2^* = b_2, \quad b_3^* = b_3, \quad b_0^* = -(b_1^* + b_2^* + b_3^*),$$

and b_1, b_2, and b_3 are given in Eq. (2.1). Here, $\lambda = (1/2)\boldsymbol{\epsilon}^\top(\boldsymbol{K}_{\beta 11} - \boldsymbol{K}_{\beta 12}\boldsymbol{K}_{\beta 22}^{-1}\boldsymbol{K}_{\beta 21})\boldsymbol{\epsilon}$, and

$$\xi = \frac{K_\phi^{-1}}{2} \sum_{t=1}^{p} (\kappa_{\phi\phi t} + 2\kappa_{\phi,\phi t})\epsilon_t^*,$$

where

$$\kappa_{\phi\phi t} = n^{-1}\mathbb{E}\left(\frac{\partial^3 \ell(\boldsymbol{\theta})}{\partial\phi^2 \partial\beta_t}\right),$$

$$\kappa_{\phi,\phi t} = n^{-1}\mathbb{E}\left(\frac{\partial \ell(\boldsymbol{\theta})}{\partial\phi}\frac{\partial^2 \ell(\boldsymbol{\theta})}{\partial\phi\partial\beta_t}\right).$$

Notice that b_1 and b_k^* ($k = 2, 3$) represent the contribution of the parameter vector $\boldsymbol{\beta}$ to the local power of the gradient test for testing the null hypothesis $\mathcal{H}_0 : \boldsymbol{\beta}_1 = \boldsymbol{\beta}_{10}$, since these expressions are only obtained over the components of $\boldsymbol{\beta}$, that is, as if ϕ were known. On the other hand, the quantity ξ, which depends on third-order mixed cumulants involving ϕ and $\boldsymbol{\beta}$, can be regarded as the contribution of the parameter ϕ to the local power of the gradient test when it is unknown, that is, when it needs to be estimated. Additionally, the contribution of the parameter ϕ to the local power of the test only appears in the coefficient b_1^* and, of course, in b_0^*.

Theorem 2.2 implies that the limiting distribution of the gradient statistic, namely a noncentral χ^2 distribution with noncentrality parameter λ, is the same regardless of whether ϕ is known or estimated from the data. Notice that ξ is the only term that involves cumulants of log-likelihood derivatives with respect to ϕ, and it decreases with the Fisher information for ϕ and vanishes if ϕ is known. By using the Bartlett identity

$$\kappa_{\phi,\phi t} = \kappa_{\phi t}^{(\phi)} - \kappa_{\phi\phi t}, \quad t = 1, \ldots, p,$$

where $\kappa_{\phi t}^{(\phi)} = \partial\kappa_{\phi t}/\partial\phi$, we have $\kappa_{\phi,\phi t} = -\kappa_{\phi\phi t}$ since the orthogonality between $\boldsymbol{\beta}$ and ϕ implies that

$$\kappa_{\phi t} = n^{-1}\mathbb{E}\left(\frac{\partial^2\ell(\boldsymbol{\theta})}{\partial\beta_t\partial\phi}\right) = 0, \quad t = 1, \ldots, p.$$

Therefore, we can write

$$\xi = -\frac{K_{\phi}^{-1}}{2}\sum_{t=1}^{p}\kappa_{\phi\phi t}\epsilon_t^{*}. \tag{2.5}$$

Theorem 2.2 has a practical application when the goal is to obtain explicit formulas for the coefficients that define the nonnull cumulative distribution function of the gradient test statistic for special models in which orthogonality holds. It suggests that the coefficients b_k^{*}s should be obtained as if the scalar orthogonal parameter ϕ were known, and the extra contribution due to the estimation of the parameter ϕ should be obtained from Eq. (2.5).

Now, let Π_{T}^{0} and Π_{T} be the local power functions (ignoring terms of order smaller than $O(n^{-1/2})$) of the gradient test when ϕ is known and when ϕ is unknown, respectively. We can then write

$$\Pi_{\mathrm{T}} - \Pi_{\mathrm{T}}^{0} = \frac{c\,\xi}{\sqrt{n}},$$

where $c = 2g_{q+2,\lambda}(x) > 0$, x represents the appropriate quantile of the reference distribution for the chosen nominal level, and $g_{\nu,\lambda}(x)$ is the probability density function of a noncentral χ^2 variate with ν degrees of freedom and noncentrality parameter λ. Therefore, the difference between the local powers can be zero, or it can increase or decrease when ϕ needs to be estimated, depending on the sign of the components of $\boldsymbol{\epsilon}$. If $\kappa_{\phi\phi t} = 0$, for $t = 1, \ldots, p$, we have $\xi = 0$ and hence the local power of the gradient test does not change when a scalar parameter, which is globally orthogonal to the remaining parameters, is included in the model specification.

If $q = p$, the null hypothesis takes the form $\mathcal{H}_0 : \boldsymbol{\beta} = \boldsymbol{\beta}_0$. Therefore, an immediate consequence of Theorem 2.2 is the following corollary.

Corollary 2.2. *Let $\boldsymbol{\theta} = (\boldsymbol{\beta}^{\top}, \phi)^{\top}$ be the parameter vector with $\boldsymbol{\beta}$ and ϕ being globally orthogonal parameters. The nonnull asymptotic expansion of* *the cumulative distribution function of the gradient statistic for testing the null hypothesis $\mathcal{H}_0 : \boldsymbol{\beta} = \boldsymbol{\beta}_0$ under a sequence of Pitman alternatives is given by*

$$\Pr(S_{\mathrm{T}} \leq x|\mathcal{H}_{\mathrm{an}}) = G_{p,\lambda}(x) + \frac{1}{\sqrt{n}} \sum_{k=0}^{3} b_k^* G_{p+2k,\lambda}(x) + O(n^{-1}),$$

where

$$b_1^* = b_1 + \xi, \quad b_2^* = b_2, \quad b_3^* = b_3, \quad b_0^* = -(b_1^* + b_2^* + b_3^*),$$

and b_1, b_2, and b_3 are given in Eq. (2.4). Also, $\lambda = (1/2)\boldsymbol{\epsilon}^{\top}\boldsymbol{K}_{\boldsymbol{\beta}}\boldsymbol{\epsilon}$ and $\xi = (1/2)K_{\phi}^{-1}\sum_{t=1}^{p}(\kappa_{\phi\phi t} + 2\kappa_{\phi,\phi t})\epsilon_t$.

If $q = p = 1$, the null hypothesis is $\mathcal{H}_0 : \beta = \beta_0$, where β_0 is a specified scalar, and hence we have the corollary.

Corollary 2.3. *Let $\boldsymbol{\theta} = (\beta, \phi)^{\top}$ be the parameter vector with β and ϕ being globally orthogonal scalar parameters. The nonnull asymptotic expansion of the cumulative distribution function of the gradient statistic for testing the null hypothesis $\mathcal{H}_0 : \beta = \beta_0$ under a sequence of Pitman alternatives is given by*

$$\Pr(S_{\mathrm{T}} \leq x|\mathcal{H}_{\mathrm{an}}) = G_{1,\lambda}(x) + \frac{1}{\sqrt{n}} \sum_{k=0}^{3} b_k^* G_{1+2k,\lambda}(x) + O(n^{-1}),$$

where $\lambda = \epsilon^2 K_{\beta}/2$, and the coefficients are $b_1^* = b_1 + \xi$, $\xi = (1/2) K_{\phi}^{-1}(\kappa_{\phi\phi\beta} + 2\kappa_{\phi,\phi\beta})\epsilon$, $b_1 = -(1/4)\kappa_{\beta\beta\beta}K_{\beta}^{-1}\epsilon + (1/2)\kappa_{\beta,\beta\beta}\epsilon^3$, $b_2^* = (1/4)\kappa_{\beta\beta\beta}K_{\beta}^{-1}\epsilon - (1/4)(\kappa_{\beta\beta\beta} + 2\kappa_{\beta,\beta\beta})\epsilon^3$, $b_3^* = (1/12)\kappa_{\beta\beta\beta}\epsilon^3$, and $b_0^* = -(b_1^* + b_2^* + b_3^*)$. Here, $\epsilon = \sqrt{n}(\beta - \beta_0)$ and K_{β} is the per observation Fisher information of β. Also, $\kappa_{\phi\phi\beta} = n^{-1}\mathbb{E}(\partial^3\ell(\boldsymbol{\theta})/\partial\phi^2\partial\beta)$, $\kappa_{\beta,\beta\beta} = n^{-1}\mathbb{E}[(\partial\ell(\boldsymbol{\theta})/\partial\beta)(\partial^2\ell(\boldsymbol{\theta})/\partial\beta^2)]$, and so on.

We now present two examples. We focus on ξ, which determines whether the local power changes if a parameter that is globally orthogonal to the remaining parameters is introduced in the model. It is evident that several other special cases could be considered.

Example 2.1 (Normal distribution). Let $x_1, \ldots, x_n$ be n independent and identically distributed random variables with probability density function in the form $f(x;\mu,\sigma^2) = (2\pi\sigma^2)^{-1/2}\exp[-(x-\mu)^2/(2\sigma^2)]$, where $x \in \mathbb{R}$, $\mu \in \mathbb{R}$, and $\sigma^2 > 0$. First, let $\mathcal{H}_0 : \mu = \mu_0$ be the null hypothesis of interest, where μ_0 is a specified scalar. In this case we have $\xi = 0$ and hence the coefficients that define the nonnull asymptotic expansion of the cumulative distribution function of the gradient statistic does not change when the parameter σ^2 needs to be estimated. Now, consider the test of the null hypothesis $\mathcal{H}_0 : \sigma^2 = \sigma_0^2$ against $\mathcal{H}_a : \sigma^2 \neq \sigma_0^2$, where σ_0^2 is a specified positive scalar. We have $\xi = -\epsilon/(2\sigma_0^2)$, where $\epsilon = \sqrt{n}(\sigma^2 - \sigma_0^2)$. Notice that the additional contribution on the local power function of the gradient test for testing $\mathcal{H}_0 : \sigma^2 = \sigma_0^2$ by considering μ unknown does not depend on this parameter.

Example 2.2 (von Mises distribution). Let $x_1,\ldots,x_n$ be n independent and identically distributed random variables with a von Mises distribution with mean direction μ, concentration parameter ϕ, and probability density function $f(x;\mu,\phi) = [2\pi I_0(\phi)]^{-1}\exp[\phi\cos(x-\mu)]$, where $0 \le x < 2\pi$, $0 \le \mu < 2\pi$, $\phi > 0$, and $I_0(\cdot)$ is the modified Bessel function of the first kind and order zero. The positive parameter ϕ measures the concentration of the distribution: as $\phi \to 0$ the von Mises distribution converges to the uniform distribution around the circumference, whereas for $\phi \to \infty$ the distribution tends to the point distribution concentrated in the mean direction. This distribution is particularly useful for the analysis of circular data. For testing $\mathcal{H}_0 : \mu = \mu_0$, where μ_0 is a specified scalar, we have $\xi = 0$ and hence when one introduces unknown concentration the coefficients that define the nonnull asymptotic expansion of the cumulative distribution function of the gradient statistic do not change. The additional contribution on the local power function of the gradient test for testing $\mathcal{H}_0 : \phi = \phi_0$, where ϕ_0 is a specified positive scalar, when the parameter μ is unknown, reduces to $\xi = \epsilon/(2\phi_0)$, where $\epsilon = \sqrt{n}(\phi - \phi_0)$. It is interesting to note that ξ does not involve the parameter μ.

2.5 ONE-PARAMETER EXPONENTIAL FAMILY

Let $x_1,\ldots,x_n$ be n independent observations with each x_l having probability density function $f(x;\theta) = \exp[t(x;\theta)]$, where θ is a scalar parameter. To test $\mathcal{H}_0 : \theta = \theta_0$, where θ_0 is a fixed known constant, the gradient statistic is given by

$$S_T = (\hat{\theta} - \theta_0)\sum_{l=1}^{n} t'(x_l;\theta_0),$$

where $\hat{\theta}$ is the MLE of θ, and $t'(x;\theta) = dt(x;\theta)/d\theta$. Under the null hypothesis, the gradient statistic has a χ_1^2 distribution asymptotically.

Let $\kappa_{\theta\theta} = \mathbb{E}[t''(x;\theta)]$, $\kappa_{\theta\theta\theta} = \mathbb{E}[t'''(x;\theta)]$, $\kappa_{\theta\theta,\theta} = \mathbb{E}[t''(x;\theta)t'(x;\theta)]$, $\kappa^{\theta,\theta} = -\kappa_{\theta\theta}^{-1}$, etc., where primes denote derivatives with respect to θ. The asymptotic expansion of the cumulative distribution function of the gradient statistic for the null hypothesis $\mathcal{H}_0 : \theta = \theta_0$ under the sequence of local alternatives $\mathcal{H}_{an} : \theta = \theta_0 + n^{-1/2}\epsilon$ is given by Eq. (2.3) with $p = 1$, $\lambda = (1/2)K\epsilon^2$, where $K = K(\theta)$ denotes the Fisher information for a single observation,

$$b_0 = \frac{\kappa_{\theta\theta\theta}\epsilon^3}{6}, \quad b_1 = -\frac{\kappa_{\theta\theta\theta}\kappa^{\theta,\theta}\epsilon - 2\kappa_{\theta,\theta\theta}\epsilon^3}{4},$$

$$b_2 = \frac{\kappa_{\theta\theta\theta}\kappa^{\theta,\theta}\epsilon - (\kappa_{\theta\theta\theta} + 2\kappa_{\theta,\theta\theta})\epsilon^3}{4}, \quad b_3 = \frac{\kappa_{\theta\theta\theta}\epsilon^3}{12},$$

and $\epsilon = \sqrt{n}(\theta - \theta_0)$.

We now specialize to the case where $f(x;\theta)$ belongs to the one-parameter exponential family. Let

$$t(x;\theta) = -\log\zeta(\theta) - \alpha(\theta)d(x) + v(x),$$

where $\alpha(\cdot)$, $\zeta(\cdot)$, $d(\cdot)$, and $v(\cdot)$ are known functions. Also, $\alpha(\cdot)$ and $\zeta(\cdot)$ are assumed to have first three continuous derivatives, with $\zeta(\cdot) > 0$, $\alpha'(\theta)$ and $\beta'(\theta)$ being different from zero for all θ in the parameter space, where $\beta(\theta) = \zeta'(\theta)/[\zeta(\theta)\alpha'(\theta)]$. We have that $K = \alpha'(\theta)\beta'(\theta)$, $\sum_{l=1}^n t(x_l;\theta) = -n[\log\zeta(\theta) + \alpha(\theta)\bar{d} - \bar{v}]$, $\sum_{l=1}^n t'(x_l;\theta) = -n\alpha'(\theta)[\beta(\theta) + \bar{d}]$, with $\bar{d} = n^{-1}\sum_{l=1}^n d(x_l)$ and $\bar{v} = n^{-1}\sum_{l=1}^n v(x_l)$. It follows that

$$S_T = n(\theta_0 - \hat{\theta})\alpha'(\theta_0)[\beta(\theta_0) + \bar{d}].$$

Let $\alpha' = \alpha'(\theta)$, $\alpha'' = \alpha''(\theta)$, $\beta' = \beta'(\theta)$, and $\beta'' = \beta''(\theta)$. It can be shown that $\kappa_{\theta\theta} = -\alpha'\beta'$, $\kappa_{\theta\theta\theta} = -(2\alpha''\beta' + \alpha'\beta'')$, $\kappa_{\theta,\theta\theta} = \alpha''\beta'$, $\kappa_{\theta,\theta,\theta} = \alpha'\beta'' - \alpha''\beta'$. The coefficients that define the local power function of the gradient test reduce to

$$b_0 = -\frac{(2\alpha''\beta' + \alpha'\beta'')\epsilon^3}{6}, \quad b_1 = \frac{\alpha''\beta'\epsilon^3}{2} + \frac{(2\alpha''\beta' + \alpha'\beta'')\epsilon}{4\alpha'\beta'},$$

$$b_2 = \frac{\alpha'\beta''\epsilon^3}{4} - \frac{(2\alpha''\beta' + \alpha'\beta'')\epsilon}{4\alpha'\beta'}, \quad b_3 = -\frac{(2\alpha''\beta' + \alpha'\beta'')\epsilon^3}{12}.$$

If $\alpha(\theta) = \theta$, $f(x;\theta)$ corresponds to the one-parameter natural exponential family. In this case, $\alpha' = 1$, $\alpha'' = 0$, and the bs simplify considerably.

All quantities in the b_ks ($k = 0, 1, 2, 3$), except ϵ, are evaluated under the null hypothesis $\mathcal{H}_0 : \theta = \theta_0$.

We now present some analytical comparisons among the local powers of the gradient, LR, Wald, and score tests for a number of distributions within the one-parameter exponential family. Let Π_i and Π_j be the local power functions, up to order $O(n^{-1/2})$, of the tests that use the statistics S_i and S_j, respectively, with $i \neq j$ and $i, j = \text{LR, W, R, T}$. We have

$$\Pi_i - \Pi_j = \frac{1}{\sqrt{n}} \sum_{k=0}^{3} (b_{jk} - b_{ik}) G_{1+2k,\lambda}(x), \tag{2.6}$$

where x is replaced by the appropriate quantile of the reference distribution for the chosen nominal level. The coefficients that define the local power functions of the LR, Wald, and score tests are given in Lemonte and Ferrari [12]. It is well known that

$$G_{m,\lambda}(x) - G_{m+2,\lambda}(x) = 2g_{m+2,\lambda}(x). \tag{2.7}$$

From Eqs. (2.6) and (2.7), we can state the following comparison among the powers of the four tests. Here, we assume that $\theta > \theta_0$; opposite inequalities hold if $\theta < \theta_0$.

1. Normal ($\theta > 0$, $\mu \in \mathbb{R}$, and $x \in \mathbb{R}$):
   - μ known: $\alpha(\theta) = (2\theta)^{-1}$, $\zeta(\theta) = \theta^{1/2}$, $d(x) = (x - \mu)^2$, and $v(x) = -(1/2)\log(2\pi)$,

$$\Pi_{\text{T}} > \Pi_{\text{R}} > \Pi_{\text{LR}} > \Pi_{\text{W}}.$$

   - θ known: $\alpha(\mu) = -\mu/\theta$, $\zeta(\mu) = \exp[\mu^2/(2\theta)]$, $d(x) = x$, and $v(x) = -[x^2 + \log(2\pi\theta)]/2$,

$$\Pi_{\text{LR}} = \Pi_{\text{W}} = \Pi_{\text{R}} = \Pi_{\text{T}}.$$

2. Inverse normal ($\theta > 0$, $\mu > 0$, and $x > 0$):
   - μ known: $\alpha(\theta) = \theta$, $\zeta(\theta) = \theta^{-1/2}$, $d(x) = (x - \mu)^2/(2\mu^2 x)$, and $v(x) = -(1/2)\log(2\pi x^3)$,

$$\Pi_{\text{LR}} > \Pi_{\text{T}} > \Pi_{\text{W}} = \Pi_{\text{R}}.$$

   - θ known: $\alpha(\mu) = \theta/(2\mu^2)$, $\zeta(\mu) = \exp(-\theta/\mu)$, $d(x) = x$, and $v(x) = -\{\theta/(2x) - \log[\theta/(2\pi x^3)]\}/2$,

$$\Pi_{\text{T}} > \Pi_{\text{R}} > \Pi_{\text{LR}} > \Pi_{\text{W}}.$$

3. Gamma ($k > 0$, k known, $\theta > 0$, and $x > 0$): $\alpha(\theta) = \theta$, $\zeta(\theta) = \theta^{-k}$, $d(x) = x$, and $v(x) = (k-1)\log x - \log \Gamma(k)$, $\Gamma(\cdot)$ is the gamma function,

$$\Pi_{\mathrm{T}} > \Pi_{\mathrm{LR}} > \Pi_{\mathrm{W}} = \Pi_{\mathrm{R}}.$$

4. Truncated extreme value ($\theta > 0$ and $x > 0$): $\alpha(\theta) = \theta^{-1}$, $\zeta(\theta) = \theta$, $d(x) = \exp(x) - 1$, and $v(x) = x$,

$$\Pi_{\mathrm{T}} > \Pi_{\mathrm{R}} > \Pi_{\mathrm{LR}} > \Pi_{\mathrm{W}}.$$

5. Pareto ($\theta > 0$, $k > 0$, k known, and $x > k$): $\alpha(\theta) = 1 + \theta$, $\zeta(\theta) = (\theta k^{\theta})^{-1}$, $d(x) = \log x$, and $v(x) = 0$,

$$\Pi_{\mathrm{T}} > \Pi_{\mathrm{LR}} > \Pi_{\mathrm{W}} = \Pi_{\mathrm{R}}.$$

6. Laplace ($\theta > 0$, $-\infty < k < \infty$, k known, and $x > 0$): $\alpha(\theta) = \theta^{-1}$, $\zeta(\theta) = 2\theta$, $d(x) = |x - k|$, and $v(x) = 0$,

$$\Pi_{\mathrm{T}} > \Pi_{\mathrm{R}} > \Pi_{\mathrm{LR}} > \Pi_{\mathrm{W}}.$$

7. Power ($\theta > 0$, $\phi > 0$, ϕ known, and $x > \phi$): $\alpha(\theta) = 1 - \theta$, $\zeta(\theta) = \theta^{-1}\phi^{\theta}$, $d(x) = \log x$, and $v(x) = 0$,

$$\Pi_{\mathrm{T}} > \Pi_{\mathrm{LR}} > \Pi_{\mathrm{W}} = \Pi_{\mathrm{R}}.$$

2.6 SYMMETRIC LINEAR REGRESSION MODELS

We say that the random variable y follows a symmetric distribution if its probability density function takes the form

$$f(y;\mu,\phi) = \frac{1}{\phi} g\left(\left(\frac{y-\mu}{\phi}\right)^2\right), \quad y \in \mathbb{R}, \tag{2.8}$$

where $\mu \in \mathbb{R}$ is a location parameter, and $\phi > 0$ is a scale parameter. The function $g : \mathbb{R} \to [0, \infty)$ is such that $\int_0^\infty g(u)du < \infty$ and $\int_0^\infty u^{-1/2} g(u)du = 1$ to guarantee that $f(\cdot;\mu,\phi)$ is a density function. We then write $y \sim S(\mu, \phi^2)$. The function $g(\cdot)$, which is independent of y, μ, and ϕ, is typically known as density generator. The probability density function of $z = (y-\mu)/\phi$ is $f(v; 0, 1) = g(v^2)$, $v \in \mathbb{R}$, that is, $z \sim S(0, 1)$. It is the standardized form of the symmetric distributions. The symmetrical family of distributions allows an extension of the normal distribution for statistical modeling of real data involving distributions with heavier and lighter tails than the ones of the normal distribution. This class of distributions is appearing with increasing frequency in the statistical literature to model several types of data containing more outlying observations than can be expected based on a normal distribution.

A number of important distributions have probability density function (2.8) and a wide range of practical applications in various fields such as engineering, biology, medicine, and economics, among others. Some special cases of Eq. (2.8) are the following: normal, Cauchy, Student-t, generalized Student-t, type I logistic, type II logistic, generalized logistic, Kotz, generalized Kotz, contaminated normal, double exponential, power exponential, and extended power distributions. These 13 distributions provide a rich source of alternative models for analyzing univariate data containing outlying observations.

In what follows, we restrict our attention to the following distributions: normal, Cauchy, Student-t, generalized Student-t, type I logistic, type II logistic, generalized logistic, and power exponential distributions. Additionally, all extra parameters will be considered as known or fixed, for example, the degrees of freedom ν for the Student-t model. The density generators are

- Normal: $g(u) = (2\pi)^{-1/2}\exp(-u/2)$;
- Cauchy: $g(u) = [\pi(1+u)]^{-1}$;
- Student-t: $g(u) = \nu^{\nu/2}B(1/2, \nu/2)^{-1}(\nu+u)^{-(\nu+1)/2}$, where $\nu > 0$ and $B(\cdot,\cdot)$ is the beta function;
- Generalized Student-t: $g(u) = s^{r/2}B(1/2, r/2)^{-1}(s+u)^{-(r+1)/2}$, where $s, r > 0$;
- Type I logistic: $g(u) = c\mathrm{e}^{-u}(1+\mathrm{e}^{-u})^{-2}$, where $c \approx 1.484300029$ is the normalizing constant obtained from $\int_0^\infty u^{-1/2}g(u)\,du = 1$;
- Type II logistic: $g(u) = \mathrm{e}^{-\sqrt{u}}(1+\mathrm{e}^{-\sqrt{u}})^{-2}$;
- Power exponential: $g(u) = c(k)\exp(-0.5u^{1/(1+k)})$, $-1 < k \leq 1$, and $c(k) = \Gamma(1+(k+1)/2)2^{1+(1+k)/2}$.

2.6.1 The Model, Estimation, and Testing

The symmetric linear regression model is defined as

$$y_l = \mu_l(\boldsymbol{\beta}) + \varepsilon_l, \quad l = 1, \ldots, n,$$

where $\mu_l = \mu_l(\boldsymbol{\beta}) = \boldsymbol{x}_l^\top\boldsymbol{\beta}$, $\boldsymbol{\beta} = (\beta_1, \ldots, \beta_p)^\top$ is a vector of unknown regression parameters, $\boldsymbol{x}_l^\top = (x_{l1}, \ldots, x_{lp})$ contains the lth observation on p covariates ($p < n$), and $\varepsilon_l \sim S(0, \phi^2)$. We have, when they exist, that $\mathbb{E}(y_l) = \mu_l$ and $\mathbb{VAR}(y_l) = k\phi^2$, where $k > 0$ is a constant that may be obtained from the expected value of the radial variable or from the derivative of the characteristic function. Note that the parameter ϕ is a kind of dispersion parameter. For example, for the Student-t model with ν degrees of freedom we have $k = \nu/(\nu - 2)$, for $\nu > 2$.

Let $\ell(\boldsymbol{\theta})$ denote the log-likelihood function for the parameter vector $\boldsymbol{\theta} = (\boldsymbol{\beta}^\top, \phi)^\top$. We have

$$\ell(\boldsymbol{\theta}) = -n \log \phi + \sum_{l=1}^{n} \log g(z_l^2),$$

where $z_l = (y_l - \mu_l)/\phi$ is the standardized lth observation. The function $\ell(\boldsymbol{\theta})$ is assumed to be regular with respect to all $\boldsymbol{\beta}$ and ϕ derivatives up to third order. The score functions for $\boldsymbol{\beta}$ and ϕ have, respectively, the forms

$$\boldsymbol{U}_{\boldsymbol{\beta}}(\boldsymbol{\theta}) = \phi^{-2}\boldsymbol{X}^\top \boldsymbol{W}(\boldsymbol{y} - \boldsymbol{\mu}), \quad U_{\phi}(\boldsymbol{\theta}) = \phi^{-1}(\phi^{-2}\boldsymbol{Q} - n),$$

where $\boldsymbol{y} = (y_1, \ldots, y_n)^\top$, $\boldsymbol{\mu} = (\mu_1, \ldots, \mu_n)^\top$, $\boldsymbol{X} = (\boldsymbol{x}_1, \ldots, \boldsymbol{x}_n)^\top$, $\boldsymbol{W} = \text{diag}\{w_1, \ldots, w_n\}$ with $w_l = -2d \log g(u)/du|_{u=z_l^2}$, and $\boldsymbol{Q} = (\boldsymbol{y} - \boldsymbol{\mu})^\top \boldsymbol{W}(\boldsymbol{y} - \boldsymbol{\mu})$. The model matrix $\boldsymbol{X}$ is assumed to be of full rank, that is, $\text{rank}(\boldsymbol{X}) = p$.

The MLE of $\boldsymbol{\theta} = (\boldsymbol{\beta}^\top, \phi)^\top$, $\hat{\boldsymbol{\theta}} = (\hat{\boldsymbol{\beta}}^\top, \hat{\phi})^\top$, can be obtained by solving simultaneously the equations $\boldsymbol{U}_{\boldsymbol{\beta}}(\hat{\boldsymbol{\theta}}) = \boldsymbol{0}_p$ and $U_{\phi}(\hat{\boldsymbol{\theta}}) = 0$, that is, the MLEs of $\boldsymbol{\beta}$ and ϕ come from the likelihood equations $\hat{\boldsymbol{\beta}} = (\boldsymbol{X}^\top \hat{\boldsymbol{W}}\boldsymbol{X})^{-1}\boldsymbol{X}^\top \hat{\boldsymbol{W}}\boldsymbol{y}$ and $\hat{\phi}^2 = n^{-1}\hat{\boldsymbol{Q}}$. Note that the weight w_l may be regarded as the contribution of the lth observation for the estimation of the parameters. For the Student-t, type I, and type II logistic distributions, w_l is given by $(\nu + 1)/(\nu + z_l^2)$, $2(1-\mathrm{e}^{-z_l^2})/(1+\mathrm{e}^{-z_l^2})$, and $(\mathrm{e}^{-|z_l|} - 1)/[|z_l|(1+\mathrm{e}^{-|z_l|})]$, respectively. For the normal model, all observations have the same weight in the estimation of $\boldsymbol{\beta}$ and ϕ. Also, for the Cauchy, Student-t, generalized Student-t, type II logistic, and power exponential ($k>0$) distributions, w_l is a decreasing function of $|z_l| = |y_l - \mu_l|/\phi$. Hence, the MLEs of $\boldsymbol{\beta}$ and ϕ are robust in the sense that the observations with large $|z_l|$ have small weight w_l. For the type I logistic distribution, the weights w_l are increasing functions of $|z_l|$. This is expected since this distribution has shorter-than-normal tails. Fig. 2.1 displays the behavior of the weights for the Student-t, power exponential, and type I logistic models. The dotted line represents the weights for the normal model.

We introduce the notation $\alpha_{r,s} = \mathbb{E}(t^{(r)}(z)z^s)$, for $r, s = 0, 1, 2, 3$, where $t(z) = \log g(z^2)$ and $t^{(r)}(z) = d^r t(z)/dz^r$. The Fisher information matrix for the parameter vector $\boldsymbol{\theta} = (\boldsymbol{\beta}^\top, \phi)^\top$ is block-diagonal and is given by

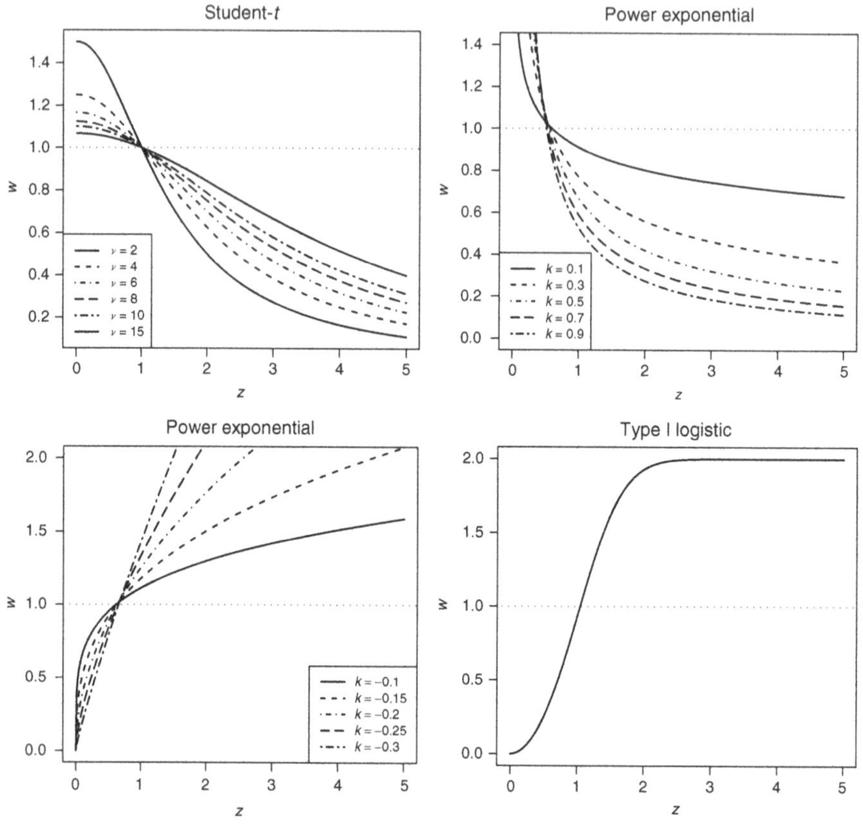

Fig. 2.1 Weights for the Student-t, power exponential, and type I logistic models.

$\boldsymbol{K}(\boldsymbol{\theta}) = \text{diag}\{\boldsymbol{K}_{\boldsymbol{\beta}}, K_{\phi}\}$, where $\boldsymbol{K}_{\boldsymbol{\beta}} = -(\alpha_{2,0}/\phi^2)\boldsymbol{X}^\top\boldsymbol{X}$ and $K_\phi = n(1 - \alpha_{2,2})/\phi^2$. Therefore, $\boldsymbol{\beta}$ and ϕ are globally orthogonal and their MLEs are asymptotically uncorrelated. The Fisher scoring method can be used to estimate $\boldsymbol{\beta}$ and ϕ simultaneously by iteratively solving the equations

$$\boldsymbol{X}^\top\boldsymbol{X}\boldsymbol{\beta}^{(m+1)} = \boldsymbol{X}^\top\boldsymbol{\zeta}^{(m)}, \quad \phi^{(m+1)} = \phi^{(m)} + \frac{n^{-1}\boldsymbol{Q}^{(m)} - \phi^{(m)2}}{\phi^{(m)}(1 - \alpha_{2,0})},$$

where $\boldsymbol{\zeta}^{(m)} = \boldsymbol{X}\boldsymbol{\beta}^{(m)} - \boldsymbol{W}^{(m)}(\boldsymbol{y} - \boldsymbol{\mu}^{(m)})/\alpha_{2,0}$ and $m = 0, 1, \ldots$. The procedure may be initialized by taking $\boldsymbol{\beta}^{(0)} = (\boldsymbol{X}^\top\boldsymbol{X})^{-1}\boldsymbol{X}^\top\boldsymbol{y}$, the least squares estimate of $\boldsymbol{\beta}$, and $\phi^{(0)} = [n^{-1}(\boldsymbol{y} - \boldsymbol{X}\boldsymbol{\beta}^{(0)})^\top(\boldsymbol{y} - \boldsymbol{X}\boldsymbol{\beta}^{(0)})]^{1/2}$. For any distribution in Eq. (2.8), the αs are easily found and satisfy standard regularity conditions which usually facilitate their calculations. The values of $\alpha_{2,0}$, $\alpha_{2,2}$, $\alpha_{3,1}$, and $\alpha_{3,3}$ are given in Table 2.1 for some symmetric distributions.

Table 2.1 Values of $\alpha_{2,0}$, $\alpha_{2,2}$, $\alpha_{3,1}$, and $\alpha_{3,3}$ for Some Symmetric Distributions

Model	$\alpha_{2,0}$	$\alpha_{2,2}$	$\alpha_{3,1}$	$\alpha_{3,3}$
Normal	-1	-1	0	0
Cauchy	$-1/2$	$1/2$	$1/2$	$-1/2$
Student-t	$-\frac{\nu+1}{\nu+3}$	$\frac{3-\nu}{\nu+3}$	$\frac{6(\nu+1)}{(\nu+3)(\nu+5)}$	$\frac{6(3\nu-5)}{(\nu+3)(\nu+5)}$
Type I logistic	-1.47724	-2.01378	-1.27916	-0.50888
Type II logistic	$-1/3$	-0.42996	$1/6$	0.64693

We are interested in testing the composite null hypothesis $\mathcal{H}_0 : \boldsymbol{\beta}_1 = \boldsymbol{\beta}_{10}$ in the class of symmetric linear regression models. This hypothesis will be tested against the alternative hypothesis $\mathcal{H}_a : \boldsymbol{\beta}_1 \neq \boldsymbol{\beta}_{10}$, where $\boldsymbol{\beta}$ is partitioned as $\boldsymbol{\beta} = (\boldsymbol{\beta}_1^\top, \boldsymbol{\beta}_2^\top)^\top$, $\boldsymbol{\beta}_1 = (\beta_1, \ldots, \beta_q)^\top$, and $\boldsymbol{\beta}_2 = (\beta_{q+1}, \ldots, \beta_p)^\top$. Here, $\boldsymbol{\beta}_{10}$ is a fixed column vector of dimension q. Let $(\hat{\boldsymbol{\beta}}_1, \hat{\boldsymbol{\beta}}_2, \hat{\phi})$ and $(\boldsymbol{\beta}_{10}, \tilde{\boldsymbol{\beta}}_2, \tilde{\phi})$ be the unrestricted and restricted MLEs of $(\boldsymbol{\beta}_1, \boldsymbol{\beta}_2, \phi)$, respectively. The gradient statistic for testing $\mathcal{H}_0 : \boldsymbol{\beta}_1 = \boldsymbol{\beta}_{10}$ can be expressed as

$$S_{\rm T} = \tilde{\phi}^{-2}(\boldsymbol{y} - \tilde{\boldsymbol{\mu}})^\top \tilde{\boldsymbol{W}} \boldsymbol{X}_1(\hat{\boldsymbol{\beta}}_1 - \boldsymbol{\beta}_{10}),$$

where the matrix $\boldsymbol{X}$ is partitioned as $\boldsymbol{X} = [\boldsymbol{X}_1 \;\; \boldsymbol{X}_2]$, $\boldsymbol{X}_1$ being $n \times q$ and $\boldsymbol{X}_2$ being $n \times (p - q)$. Here, tildes indicate evaluation at the restricted MLEs. The limiting distribution of the gradient statistic under $\mathcal{H}_0$ is χ^2_q.

2.6.2 Nonnull Distribution and Finite Sample Performance

In this section, we assume the local alternative hypothesis $\mathcal{H}_{\rm an} : \boldsymbol{\beta}_1 = \boldsymbol{\beta}_{10} + n^{-1/2}\boldsymbol{\epsilon}$, where $\boldsymbol{\epsilon} = \sqrt{n}(\boldsymbol{\beta}_1 - \boldsymbol{\beta}_{10})$. We can show that the asymptotic expansion for the nonnull cumulative distribution function of the gradient statistic is given by

$$\Pr(S_{\rm T} \leq x | \mathcal{H}_{\rm an}) = G_{q,\lambda}(x) + O(n^{-1}),$$

where $\lambda = -[\alpha_{2,0}/(2\phi^2)]\boldsymbol{\epsilon}^\top[\boldsymbol{X}_1^\top \boldsymbol{X}_1 - \boldsymbol{X}_1^\top \boldsymbol{X}_2(\boldsymbol{X}_2^\top \boldsymbol{X}_2)^{-1}\boldsymbol{X}_2^\top \boldsymbol{X}_1]\boldsymbol{\epsilon}$. The above result is very interesting. It means that the gradient test shares the same size and local power up to an error of order $O(n^{-1})$ for testing the null hypothesis $\mathcal{H}_0 : \boldsymbol{\beta}_1 = \boldsymbol{\beta}_{10}$ in the class of symmetric linear regression models. It can be shown that the LR, score, and Wald tests also share the same size and local power up to an error of order $O(n^{-1})$; see Lemonte [16]. To this order of approximation, the null distribution (ie, the distribution under the null hypothesis) of the LR, Wald, score, and gradient statistics is χ^2_q. Hence, if the sample size is large, all tests could be recommended, since their type I error probabilities do not significantly deviate from the true nominal level and their local powers are approximately equal. The natural question is how these tests perform when the sample size is small or of moderate size, and which one is the most reliable in testing hypotheses in this class of models. Next, we shall use Monte Carlo simulations to shed some light on this issue.

We conduct Monte Carlo simulations to compare the performance of the LR (S_{LR}), Wald (S_{W}), score (S_{R}), and gradient (S_{T}) tests in small- and moderate-sized samples. We consider the linear regression model

$$y_l = \beta_1 x_{l1} + \beta_2 x_{l2} + \cdots + \beta_p x_{lp} + \varepsilon_l,$$

where $x_{l1} = 1$ and $\varepsilon_l \sim S(0, \phi^2)$, $l = 1, \ldots, n$. The response is generated from a normal and Student-t (with $\nu = 4$) distributions. The covariate values were selected as random draws from the uniform $\mathcal{U}(0, 1)$ distribution and for fixed n those values were kept constant throughout the experiment. The number of Monte Carlo replications was 15,000, the nominal levels of the tests were $\gamma = 10\%$ and 5%, and all simulations were performed using the Ox matrix programming language (http://www.doornik.com).

At the outset, the null hypothesis is $\mathcal{H}_0 : \beta_1 = \beta_2 = 0$, which is tested against a two-sided alternative. The sample size is $n = 30$, $\phi = 2$, and $p = 3, \ldots, 6$. The values of the response were generated using $\beta_3 = \cdots = \beta_p = 1$. The null rejection rates of the four tests are presented in Table 2.2. It is evident that the LR and Wald tests are markedly liberal, rejecting the null hypothesis more frequently than expected based on the selected nominal levels, more so as the number of regressors increases. The score and gradient tests are also liberal in most of the cases, but much less size distorted than the LR and Wald tests in all cases. For instance, under the Student-t model, when $p = 5$ and $\gamma = 5\%$, the null rejection rates are 8.59% (S_{LR}), 12.25% (S_{W}), 6.27% (S_{R}), and 6.25% (S_{T}). Notice that the null rejection rates of the score and gradient tests are exactly the same for the normal model in all cases considered. It is noticeable that the score and gradient tests are much less liberal than the LR and Wald tests. The score and gradient tests are slightly conservative in some cases. Additionally, the Wald test is much more liberal than the other tests.

Table 2.3 reports results for $\phi = 2$, $p = 4$, and sample sizes ranging from 15 to 120. As expected, the null rejection rates of all tests approach the corresponding nominal levels as the sample size grows. Again, the score and gradient tests present the best performances.

Table 2.2 Null Rejection Rates (%): $\phi = 2$ and $n = 30$

p	$\gamma = 10\%$				$\gamma = 5\%$			
	S_{LR}	S_W	S_R	S_T	S_{LR}	S_W	S_R	S_T
Normal model								
3	12.43	14.48	10.37	10.37	6.32	8.41	4.44	4.44
4	13.39	15.45	11.24	11.24	7.07	9.04	5.21	5.21
5	14.57	16.65	12.37	12.37	8.25	10.07	6.14	6.14
6	15.91	17.93	13.51	13.51	8.86	11.11	6.59	6.59
Student-t model								
3	12.91	15.97	10.88	10.49	6.89	9.63	5.31	4.93
4	13.87	17.34	11.61	11.69	7.73	11.06	5.73	5.51
5	15.11	19.44	12.27	12.59	8.59	12.25	6.27	6.25
6	16.78	21.73	13.34	14.17	10.07	14.50	7.03	7.35

Table 2.3 Null Rejection Rates (%): $\phi = 2, p = 4$, and Different Sample Sizes

n	$\gamma = 10\%$				$\gamma = 5\%$			
	S_{LR}	S_W	S_R	S_T	S_{LR}	S_W	S_R	S_T
Normal model								
15	18.58	22.90	13.24	13.24	11.03	15.75	6.05	6.05
25	14.64	17.13	12.02	12.02	7.99	10.55	5.37	5.37
40	12.39	13.81	10.80	10.80	6.41	7.71	5.26	5.26
80	11.06	11.75	10.35	10.35	5.84	6.39	5.11	5.11
120	10.81	11.30	10.23	10.23	5.40	5.81	4.99	4.99
Student-t model								
15	20.10	27.85	13.45	13.48	12.43	20.51	6.13	5.87
25	15.15	19.27	11.95	11.99	8.65	12.75	5.83	5.75
40	13.03	15.76	11.05	11.20	6.95	9.34	5.58	5.38
80	11.81	13.09	11.01	11.00	6.10	7.11	5.39	5.43
120	10.93	11.79	10.43	10.39	5.71	6.14	5.26	5.30

A second simulation study was performed to evaluate the impact of the dimension of the null hypothesis (q) on the different tests. We consider $p = 7$, $\phi = 2$, and sample size $n = 45$. We consider the settings: (i) $\mathcal{H}_0 : \beta_1 = 0$ ($q = 1$); (ii) $\mathcal{H}_0 : \beta_1 = \beta_2 = 0$ ($q = 2$); (iii) $\mathcal{H}_0 : \beta_1 = \beta_2 = \beta_3 = 0$ ($q = 3$); (iv) $\mathcal{H}_0 : \beta_1 = \beta_2 = \beta_3 = \beta_4 = 0$ ($q = 4$); and (v) $\mathcal{H}_0 : \beta_1 = \beta_2 = \beta_3 = \beta_4 = \beta_5 = 0$ ($q = 5$). The null rejection rates of the tests are presented in Table 2.4. Note that the LR and Wald tests become liberal as the

Table 2.4 Null Rejection Rates (%): $\phi = 2, p = 7$, and $n = 45$

q	$\gamma = 10\%$				$\gamma = 5\%$			
	S_{LR}	S_W	S_R	S_T	S_{LR}	S_W	S_R	S_T
Normal model								
1	13.30	13.86	12.70	12.70	7.61	8.13	7.01	7.01
2	14.30	15.66	12.85	12.85	8.05	9.40	6.85	6.85
3	14.94	17.12	12.55	12.55	8.27	10.47	6.19	6.19
4	15.05	18.33	11.59	11.59	8.36	11.28	5.58	5.58
5	15.44	19.75	10.84	10.84	8.42	12.43	4.91	4.91
Student-*t* model								
1	13.97	15.95	12.45	13.31	7.87	9.43	6.59	7.11
2	15.27	18.55	12.71	13.67	8.69	11.81	6.54	7.01
3	16.01	20.85	12.39	13.01	8.76	13.34	6.20	6.33
4	15.95	22.77	12.03	12.05	9.23	15.15	5.79	5.93
5	16.07	23.98	11.11	11.05	9.31	16.11	5.25	5.06

dimension of the null hypothesis increases, whereas the score and gradient tests become conservative. Additionally, the score and gradient tests are less size distorted than the LR and Wald tests in all cases considered; that is, the score and gradient tests present the best performances.

We now turn to the finite-sample power properties of the four tests. The simulation results above show that the tests have different sizes when one uses their asymptotic χ^2 distribution in small and moderate-sized samples. In evaluating the power of these tests, it is important to ensure that they all have the correct size under the null hypothesis. To overcome this difficulty, we used 500,000 Monte Carlo simulated samples, drawn under the null hypothesis, to estimate the exact critical value of each test for the chosen nominal level. We set $n = 30, p = 3, \phi = 2$, and $\gamma = 10\%$. For the power simulations we compute the rejection rates under the alternative hypothesis $\beta_1 = \beta_2 = \delta$, for δ ranging from -5.0 to 5.0. Fig. 2.2 shows that the power curves of the four tests are indistinguishable from each other. As expected, the powers of the tests approach 1 as $|\delta|$ grows. Power simulations carried out for other values of n, p, and γ showed a similar pattern.

Overall, in small to moderate-sized samples, the best performing tests are the score and gradient tests. They are less size distorted than the other two and are as powerful as the others. Hence, these tests may be recommended

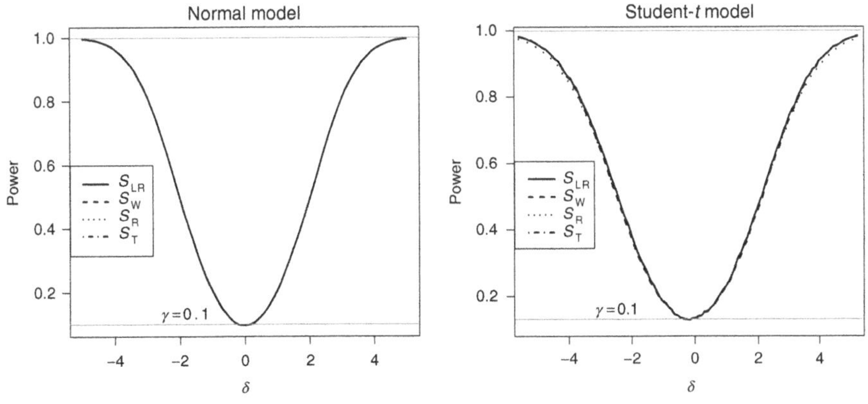

Fig. 2.2 Power of four tests: $n = 30$, $p = 3$, $\phi = 2$, and $\gamma = 10\%$.

for testing hypotheses on the regression parameters in the class of symmetric linear regression models. The gradient test has a slight advantage over the score test because the gradient statistic is simpler to calculate than the score statistic for testing a subset of regression parameters. In particular, no matrix needs to be inverted; see Lemonte [16, Section 2].

2.6.3 Tests for the Parameter ϕ

In this section, we present the asymptotic expansion for the nonnull cumulative distribution function of the gradient statistic for testing the parameter ϕ in symmetric linear regression models. We are interested in testing the null hypothesis $\mathcal{H}_0 : \phi = \phi_0$ against a two-sided alternative hypothesis $\mathcal{H}_a : \phi \neq \phi_0$, where ϕ_0 is a positive specified value for ϕ. Here, $\boldsymbol{\beta}$ acts as a nuisance parameter vector. Let $\tilde{\boldsymbol{\beta}}$ be the MLE of $\boldsymbol{\beta}$ under the null hypothesis $\mathcal{H}_0 : \phi = \phi_0$, and $\bar{m}(\boldsymbol{\beta}, \phi) = n^{-1}\sum_{l=1}^{n} w_l z_l^2$. The gradient statistic for testing $\phi = \phi_0$ takes the form

$$S_{\mathrm{T}} = n[\phi_0^{-2}\bar{m}(\tilde{\boldsymbol{\beta}}, \phi_0) - 1]\left(\frac{\hat{\phi} - \phi_0}{\phi_0}\right).$$

Under the null hypothesis, this statistic has a χ_1^2 distribution asymptotically.

The nonnull asymptotic distribution function of S_{T} for testing $\mathcal{H}_0 : \phi = \phi_0$ in symmetric linear regression models under the local alternative $\mathcal{H}_{\mathrm{an}} : \phi = \phi_0 + n^{-1/2}\epsilon$ is given by Eq. (2.2) with $q = 1$, noncentrality parameter

$\lambda = (1-\alpha_{2,2})\epsilon^2/(2\phi^2)$ and, after some algebra, the coefficients b_ks can be expressed as

$$b_1 = \frac{(\alpha_{3,3}+4\alpha_{2,2}-2)\epsilon^3}{2\phi^3} - \frac{p(\alpha_{3,1}+2\alpha_{2,0})\epsilon}{2\phi\alpha_{2,0}} + \frac{(\alpha_{3,3}+6\alpha_{2,2}-4)\epsilon}{4\phi(1-\alpha_{2,2})},$$

$$b_2 = -\frac{(\alpha_{3,3}+2\alpha_{2,2})\epsilon^3}{4\phi^3} - \frac{(\alpha_{3,3}+6\alpha_{2,2}-4)\epsilon}{4\phi(1-\alpha_{2,2})},$$

$$b_3 = -\frac{(\alpha_{3,3}+6\alpha_{2,2}-4)\epsilon^3}{12\phi^3}, \quad b_0 = -(b_1+b_2+b_3),$$

where $\epsilon = \sqrt{n}(\phi-\phi_0)$. It should be noticed that the above expressions depend on the model only through ϕ and the rank of the model matrix $\boldsymbol{X}$. They do not involve the unknown parameter vector $\boldsymbol{\beta}$. All quantities in the b_ks ($k = 0, 1, 2, 3$), except ϵ, are evaluated under the null hypothesis $\mathcal{H}_0 : \phi = \phi_0$.

In the following, we present an analytical comparison among the local powers of the LR, Wald, score, and gradient tests for testing the null hypothesis $\mathcal{H}_0 : \phi = \phi_0$. Let Π_i and Π_j be the power functions, up to order $O(n^{-1/2})$, of the tests that use the statistics S_i and S_j, respectively, with $i \neq j$ and $i, j =$ LR, W, R, T. The coefficients that define the local power functions of the LR, Wald, and score tests are given in Lemonte [16]. From Eqs. (2.6) and (2.7), and after some algebra, we can write

$$\Pi_{\mathrm{LR}} - \Pi_{\mathrm{W}} = -n^{-1/2}(\alpha_{3,3}+6\alpha_{2,2}-4)\left[\frac{\epsilon g_{5,\lambda}(x)}{\phi(1-\alpha_{2,2})} + \frac{n\epsilon^3 g_{7,\lambda}(x)}{6\phi^3}\right],$$

$$\Pi_{\mathrm{LR}} - \Pi_{\mathrm{R}} = -n^{-1/2}(1-\alpha_{3,3}-3\alpha_{2,2})\left[\frac{2\epsilon g_{5,\lambda}(x)}{\phi(1-\alpha_{2,2})} + \frac{2n\epsilon^3 g_{7,\lambda}(x)}{3\phi^3}\right],$$

$$\Pi_{\mathrm{LR}} - \Pi_{\mathrm{T}} = n^{-1/2}(\alpha_{3,3}+6\alpha_{2,2}-4)\left[\frac{\epsilon g_{5,\lambda}(x)}{2\phi(1-\alpha_{2,2})} + \frac{n\epsilon^3 g_{7,\lambda}(x)}{6\phi^3}\right],$$

$$\Pi_{\mathrm{W}} - \Pi_{\mathrm{R}} = n^{-1/2}(\alpha_{3,3}+4\alpha_{2,2}-2)\left[\frac{3\epsilon g_{5,\lambda}(x)}{\phi(1-\alpha_{2,2})} + \frac{n\epsilon^3 g_{7,\lambda}(x)}{\phi^3}\right],$$

$$\Pi_{\mathrm{W}} - \Pi_{\mathrm{T}} = n^{-1/2}(\alpha_{3,3}+6\alpha_{2,2}-4)\left[\frac{3\epsilon g_{5,\lambda}(x)}{2\phi(1-\alpha_{2,2})} + \frac{n\epsilon^3 g_{7,\lambda}(x)}{2\phi^3}\right],$$

$$\Pi_{\mathrm{R}} - \Pi_{\mathrm{T}} = -n^{-1/2}(\alpha_{3,3}+2\alpha_{2,2})\left[\frac{3\epsilon g_{5,\lambda}(x)}{2\phi(1-\alpha_{2,2})} + \frac{n\epsilon^3 g_{7,\lambda}(x)}{2\phi^3}\right].$$

From the above expressions, we arrive at the following conclusions. Here, we assume that $\phi > \phi_0$; opposite inequalities hold if $\phi < \phi_0$.

1. Normal:

$$\Pi_{\mathrm{R}} > \Pi_{\mathrm{T}} > \Pi_{\mathrm{LR}} > \Pi_{\mathrm{W}}.$$

2. Cauchy:

$$\Pi_{\mathrm{T}} > \Pi_{\mathrm{LR}} = \Pi_{\mathrm{R}} > \Pi_{\mathrm{W}}.$$

3. Student-t:
   - $0 < \nu < 1$: $\Pi_{\mathrm{T}} > \Pi_{\mathrm{LR}} > \Pi_{\mathrm{R}} > \Pi_{\mathrm{W}}$.
   - $\nu = 1$: $\Pi_{\mathrm{T}} > \Pi_{\mathrm{LR}} = \Pi_{\mathrm{R}} > \Pi_{\mathrm{W}}$.
   - $1 < \nu < 7$: $\Pi_{\mathrm{T}} > \Pi_{\mathrm{R}} > \Pi_{\mathrm{LR}} > \Pi_{\mathrm{W}}$.
   - $\nu = 7$: $\Pi_{\mathrm{R}} = \Pi_{\mathrm{T}} > \Pi_{\mathrm{LR}} > \Pi_{\mathrm{W}}$.
   - $\nu > 7$: $\Pi_{\mathrm{R}} > \Pi_{\mathrm{T}} > \Pi_{\mathrm{LR}} > \Pi_{\mathrm{W}}$.

   Note that the power inequalities of the tests depend also on the value of the degrees of freedom.
4. Generalized Student-t: replace ν with r in 3.
5. Type I logistic:

$$\Pi_{\mathrm{R}} > \Pi_{\mathrm{T}} > \Pi_{\mathrm{LR}} > \Pi_{\mathrm{W}}.$$

6. Type II logistic:

$$\Pi_{\mathrm{R}} > \Pi_{\mathrm{T}} > \Pi_{\mathrm{LR}} > \Pi_{\mathrm{W}}.$$

7. Power exponential:
   - $|k| < 1$: $\Pi_{\mathrm{R}} > \Pi_{\mathrm{T}} > \Pi_{\mathrm{LR}} > \Pi_{\mathrm{W}}$.
   - $k = 1$: $\Pi_{\mathrm{R}} = \Pi_{\mathrm{T}} > \Pi_{\mathrm{LR}} > \Pi_{\mathrm{W}}$.

2.7 GLM WITH DISPERSION COVARIATES

The class of generalized linear models (GLMs) with dispersion covariates, where the dispersion parameter of the response is a function of extra covariates, is a class of models that allows the simultaneous modeling of the mean and the dispersion by making use of the GLM framework; see, for example, McCullagh and Nelder [17].

Suppose that the univariate variables $y_1, \ldots, y_n$ are independent and each y_l has a probability density function in the following family of distributions

$$f(y; \theta_l, \phi_l) = \exp\{\phi_l[y\theta_l - b(\theta_l) + a_1(y)] + a_2(y) + c(\phi_l)\}, \qquad (2.9)$$

where $b(\cdot)$, $a_1(\cdot)$, $a_2(\cdot)$, and $c(\cdot)$ are known appropriate functions. The mean and variance are $\mathbb{E}(y_l) = \mu_l = db(\theta_l)/d\theta_l$ and $\mathbb{VAR}(y_l) = \phi_l^{-1} V_l$, where $V_l = d\mu_l/d\theta_l$ is called the variance function and $\theta_l = q(\mu_l) = \int V_l^{-1}\, d\mu_l$

is a known one-to-one function of μ_l for $l = 1, \ldots, n$. The choice of the variance function V_l as a function of μ_l determines $q(\mu_l)$. We have $V_l = 1\,[q(\mu_l) = \mu_l]$, $V_l = \mu_l^2\,[q(\mu_l) = -1/\mu_l]$, and $V_l = \mu_l^3\,[q(\mu_l) = -1/(2\mu_l^2)]$ for the normal, gamma, and inverse Gaussian distributions, respectively. The parameters θ_l and $\phi_l > 0$ in Eq. (2.9) are called the canonical and precision parameters, respectively, and the inverse of ϕ_l, ϕ_l^{-1}, is the dispersion parameter of the distribution.

We assume that both parameters μ_l and ϕ_l vary across observations through regression structures which are parameterized as $\mu_l = \mu_l(\boldsymbol{\beta})$ and $\phi_l = \phi_l(\boldsymbol{\gamma})$, where $\boldsymbol{\beta} = (\beta_1, \ldots, \beta_p)^\top$ and $\boldsymbol{\gamma} = (\gamma_1, \ldots, \gamma_q)^\top$. In classical GLMs the precision parameter is constant although possibly unknown. The usual systematic component for the mean is $d(\mu_l) = \eta_l = \boldsymbol{x}_l^\top \boldsymbol{\beta}$, where $d(\cdot)$ is the mean link function and $\boldsymbol{x}_l^\top = (x_{l1}, \ldots, x_{lp})$ is a vector of known explanatory variables, that is, $d(\boldsymbol{\mu}) = \boldsymbol{\eta} = \boldsymbol{X}\boldsymbol{\beta}$, $\boldsymbol{\mu} = (\mu_1, \ldots, \mu_n)^\top$, $\boldsymbol{\eta} = (\eta_1, \ldots, \eta_n)^\top$, and $\boldsymbol{X} = (\boldsymbol{x}_1, \ldots, \boldsymbol{x}_n)^\top$ is a specified $n \times p$ matrix of full rank $p < n$. Analogously, we consider for the precision parameter the systematic component $h(\phi_l) = \tau_l = \boldsymbol{s}_l^\top \boldsymbol{\gamma}$, where $h(\cdot)$ is the dispersion link function and $\boldsymbol{s}_l^\top = (s_{l1}, \ldots, s_{lq})$ is a vector of known variables, that is, the linear structure in $h(\phi_l)$ measures the dispersion for the lth observation. We then have $h(\boldsymbol{\phi}) = \boldsymbol{\tau} = \boldsymbol{S}\boldsymbol{\gamma}$, where $\boldsymbol{\phi} = (\phi_1, \ldots, \phi_n)^\top$, $\boldsymbol{\tau} = (\tau_1, \ldots, \tau_n)^\top$, and $\boldsymbol{S} = (\boldsymbol{s}_1, \ldots, \boldsymbol{s}_n)^\top$ is a specified $n \times q$ matrix of full rank $q < n$. The dispersion covariates in $\boldsymbol{S}$ are commonly, but not necessary, a subset of the regression covariates in $\boldsymbol{X}$. It is assumed that $\boldsymbol{\beta}$ is functionally independent of $\boldsymbol{\gamma}$. Both $d(\cdot)$ and $h(\cdot)$ are known one-to-one continuously twice differentiable functions.

Since $\boldsymbol{\beta}$ and $\boldsymbol{\gamma}$ represent the effects of the explanatory variables on the mean response and dispersion parameter, respectively, we are interested in simultaneously estimating these parameters. Let $\ell(\boldsymbol{\beta}, \boldsymbol{\gamma})$ be the log-likelihood function for a given GLM with dispersion covariates. This function is assumed to be regular with respect to all $\boldsymbol{\beta}$ and $\boldsymbol{\gamma}$ derivatives up to third order. From now on, we shall use the following notation: $\phi_{il} = d^i\phi_l/d\tau_l^i$ and $c_{il} = d^i c(\phi_l)/d\phi_l^i$, for $i = 1, 2, 3$ and $l = 1, \ldots, n$. The score function is given by $\boldsymbol{U}(\boldsymbol{\beta}, \boldsymbol{\gamma}) = (\boldsymbol{U}_{\boldsymbol{\beta}}^\top, \boldsymbol{U}_{\boldsymbol{\gamma}}^\top)^\top$, $\boldsymbol{U}_{\boldsymbol{\beta}} = \boldsymbol{U}_{\boldsymbol{\beta}}(\boldsymbol{\beta}, \boldsymbol{\gamma}) = \boldsymbol{X}^\top \boldsymbol{\Phi} \boldsymbol{W}^{1/2} \boldsymbol{V}^{-1/2}(\boldsymbol{y} - \boldsymbol{\mu})$, and $\boldsymbol{U}_{\boldsymbol{\gamma}} = \boldsymbol{U}_{\boldsymbol{\gamma}}(\boldsymbol{\beta}, \boldsymbol{\gamma}) = \boldsymbol{S}^\top \boldsymbol{\Phi}_1 \boldsymbol{v}$, where $\boldsymbol{y} = (y_1, \ldots, y_n)^\top$, $\boldsymbol{\Phi} = \mathrm{diag}\{\phi_1, \ldots, \phi_n\}$, $\boldsymbol{V} = \mathrm{diag}\{V_1, \ldots, V_n\}$, $\boldsymbol{\Phi}_1 = \mathrm{diag}\{\phi_{11}, \ldots, \phi_{1n}\}$, $\boldsymbol{W} = \mathrm{diag}\{w_1, \ldots, w_n\}$ with $w_l = V_l^{-1}(d\mu_l/d\eta_l)^2$, and $\boldsymbol{v} = (v_1, \ldots, v_n)^\top$ with $v_l = y_l\theta_l - b(\theta_l) + a_1(y_l) + c_{1l}$. The Fisher information matrix for the parameters $\boldsymbol{\beta}$ and $\boldsymbol{\gamma}$ is block-diagonal and is given by $\boldsymbol{K}(\boldsymbol{\beta}, \boldsymbol{\gamma}) = \mathrm{diag}\{\boldsymbol{K}_{\boldsymbol{\beta}}, \boldsymbol{K}_{\boldsymbol{\gamma}}\}$, where $\boldsymbol{K}_{\boldsymbol{\beta}} = \boldsymbol{X}^\top \boldsymbol{\Phi} \boldsymbol{W} \boldsymbol{X}$, $\boldsymbol{K}_{\boldsymbol{\gamma}} = \boldsymbol{S}^\top(-\boldsymbol{D}_2 \boldsymbol{\Phi}_1^{(2)})\boldsymbol{S}$,

$\boldsymbol{D}_2 = \text{diag}\{c_{21}, \ldots, c_{2n}\}$, and $\boldsymbol{\Phi}_1^{(2)} = \boldsymbol{\Phi}_1 \odot \boldsymbol{\Phi}_1$. Hereafter, "$\odot$" denotes the Hadamard (direct) product of matrices. Note that $\boldsymbol{\beta}$ and $\boldsymbol{\gamma}$ are globally orthogonal and their MLEs $\hat{\boldsymbol{\beta}}$ and $\hat{\boldsymbol{\gamma}}$ are asymptotically uncorrelated. The Fisher scoring method can be used to estimate $\boldsymbol{\beta}$ and $\boldsymbol{\gamma}$ simultaneously by iteratively solving the following equations:

$$\begin{aligned} \boldsymbol{X}^\top \boldsymbol{\Phi}^{(m)} \boldsymbol{W}^{(m)} \boldsymbol{X} \boldsymbol{\beta}^{(m+1)} &= \boldsymbol{X}^\top \boldsymbol{\Phi}^{(m)} \boldsymbol{W}^{(m)} \boldsymbol{\zeta}_1^{(m)}, \\ \boldsymbol{S}^\top (-\boldsymbol{D}_2^{(m)} \boldsymbol{\Phi}_1^{(2)(m)}) \boldsymbol{S} \boldsymbol{\gamma}^{(m+1)} &= \boldsymbol{S}^\top (-\boldsymbol{D}_2^{(m)} \boldsymbol{\Phi}_1^{(2)(m)}) \boldsymbol{\zeta}_2^{(m)}, \end{aligned} \tag{2.10}$$

where $\boldsymbol{\zeta}_1^{(m)} = \boldsymbol{X}\boldsymbol{\beta}^{(m)} + (\boldsymbol{W}^{(m)})^{-1/2} (\boldsymbol{V}^{(m)})^{-1/2} (\boldsymbol{y} - \boldsymbol{\mu}^{(m)})$, $\boldsymbol{\zeta}_2^{(m)} = \boldsymbol{S}\boldsymbol{\gamma}^{(m)} + (-\boldsymbol{D}_2^{(m)} \boldsymbol{\Phi}_1^{(m)})^{-1} \boldsymbol{v}^{(m)}$ and $m = 0, 1, \ldots$. Eq. (2.10) shows that any software with a weighted regression routine can be used to evaluate the MLEs $\hat{\boldsymbol{\beta}}$ and $\hat{\boldsymbol{\gamma}}$.

First, we are interested in testing the mean effects, that is, we consider the composite null hypothesis $\mathcal{H}_0^1 : \boldsymbol{\beta}_1 = \boldsymbol{\beta}_{10}$, which will be tested against the alternative hypothesis $\mathcal{H}_a^1 : \boldsymbol{\beta}_1 \neq \boldsymbol{\beta}_{10}$, where $\boldsymbol{\beta}$ is partitioned as $\boldsymbol{\beta} = (\boldsymbol{\beta}_1^\top, \boldsymbol{\beta}_2^\top)^\top$, $\boldsymbol{\beta}_1 = (\beta_1, \ldots, \beta_{p_1})^\top$, and $\boldsymbol{\beta}_2 = (\beta_{p_1+1}, \ldots, \beta_p)^\top$. Here, $\boldsymbol{\beta}_{10}$ is a fixed column vector of dimension p_1, and $\boldsymbol{\beta}_2$ and $\boldsymbol{\gamma}$ act as nuisance parameter vectors. Let $(\hat{\boldsymbol{\beta}}_1, \hat{\boldsymbol{\beta}}_2, \hat{\boldsymbol{\gamma}})$ and $(\boldsymbol{\beta}_{10}, \tilde{\boldsymbol{\beta}}_2, \tilde{\boldsymbol{\gamma}})$ be the unrestricted and restricted (obtained under $\mathcal{H}_0^1$) MLEs of $(\boldsymbol{\beta}_1, \boldsymbol{\beta}_2, \boldsymbol{\gamma})$, respectively. The gradient statistic for testing $\mathcal{H}_0^1$ can be expressed as

$$S_{\mathrm{T}}' = \tilde{z}_\beta^\top \boldsymbol{X}_1 (\hat{\boldsymbol{\beta}}_1 - \boldsymbol{\beta}_{10}),$$

where the matrix $\boldsymbol{X}$ is partitioned as $\boldsymbol{X} = [\boldsymbol{X}_1 \;\; \boldsymbol{X}_2]$, $\boldsymbol{X}_1$ being $n \times p_1$ and $\boldsymbol{X}_2$ being $n \times (p - p_1)$, $\tilde{z}_\beta = \tilde{\boldsymbol{\Phi}} \tilde{\boldsymbol{W}}^{1/2} \tilde{\boldsymbol{V}}^{-1/2} (\boldsymbol{y} - \tilde{\boldsymbol{\mu}})$, and tildes indicate quantities available at the restricted MLEs. The limiting distribution of this statistic under $\mathcal{H}_0^1$ is $\chi^2_{p_1}$.

Next, the null hypothesis to be considered is $\mathcal{H}_0^2 : \boldsymbol{\gamma}_1 = \boldsymbol{\gamma}_{10}$, that is, we now are interested in testing the dispersion effects. This null hypothesis will be tested against the alternative hypothesis $\mathcal{H}_a^2 : \boldsymbol{\gamma}_1 \neq \boldsymbol{\gamma}_{10}$, where $\boldsymbol{\gamma}$ is partitioned as $\boldsymbol{\gamma} = (\boldsymbol{\gamma}_1^\top, \boldsymbol{\gamma}_2^\top)^\top$, $\boldsymbol{\gamma}_1 = (\gamma_1, \ldots, \gamma_{q_1})^\top$, $\boldsymbol{\gamma}_2 = (\beta_{q_1+1}, \ldots, \gamma_q)^\top$, $\boldsymbol{\gamma}_{10}$ is a fixed column vector of dimension q_1, and $\boldsymbol{\beta}$ and $\boldsymbol{\gamma}_2$ are nuisance parameter vectors. Let $(\tilde{\boldsymbol{\beta}}, \boldsymbol{\gamma}_{10}, \boldsymbol{\gamma}_2)$ be the restricted (obtained under $\mathcal{H}_0^2$) MLE of $(\boldsymbol{\beta}, \boldsymbol{\gamma}_1, \boldsymbol{\gamma}_2)$. The gradient statistic for testing $\mathcal{H}_0^2$ takes the form

$$S_{\mathrm{T}}'' = \tilde{\boldsymbol{v}}^\top \tilde{\boldsymbol{\Phi}}_1 \boldsymbol{S}_1 (\hat{\boldsymbol{\gamma}}_1 - \boldsymbol{\gamma}_{10}),$$

where the matrix $\boldsymbol{S}$ is partitioned as $\boldsymbol{S} = [\boldsymbol{S}_1 \;\; \boldsymbol{S}_2]$, $\boldsymbol{S}_1$ being $n \times q_1$ and $\boldsymbol{S}_2$ being $n \times (q - q_1)$. Again, tildes indicate quantities available at the restricted MLEs. The limiting distribution of this statistic under $\mathcal{H}_0^2$ is $\chi^2_{q_1}$.

If $h(\phi_l) = \phi_l = \phi$ for all $l = 1, \ldots, n$, then the above expression for testing $\mathcal{H}_0^2 : \phi = \phi_0$, where ϕ_0 is a positive specified value for ϕ, reduces to

$$S_{\mathrm{T}}'' = n[\dot{c}(\phi_0) - \dot{c}(\hat{\phi})](\hat{\phi} - \phi_0),$$

where $\hat{\phi}$ is the MLE of ϕ, $\dot{c}(\phi) = dc(\phi)/d\phi$, and $\ddot{c}(\phi) = d\dot{c}(\phi)/d\phi$. Under the null hypothesis, this statistic has a χ_1^2 distribution asymptotically. For example, we have $c(\phi) = (1/2)\log\phi$ for the normal and inverse Gaussian models, which yields

$$S_{\mathrm{T}}'' = \frac{n}{2}\frac{(\hat{\phi} - \phi_0)^2}{\phi_0\hat{\phi}}.$$

For the gamma model, we have that

$$S_{\mathrm{T}}'' = n(\hat{\phi} - \phi_0)\left[\log\left(\frac{\hat{\phi}}{\phi_0}\right) + \psi(\hat{\phi}) - \psi(\phi_0)\right],$$

where $\psi(\cdot)$ denotes the digamma function.

2.7.1 Nonnull Distribution Under $\mathcal{H}_{\mathrm{an}}^1 : \boldsymbol{\beta}_1 = \boldsymbol{\beta}_{10} + \boldsymbol{\epsilon}_1$

We present in this section the expression for the nonnull asymptotic expansion up to order $O(n^{-1/2})$ for the nonnull cumulative distribution function of the gradient statistic for testing a subset of regression parameters in GLMs with dispersion covariates. We shall assume the local alternative hypothesis $\mathcal{H}_{\mathrm{an}}^1 : \boldsymbol{\beta}_1 = \boldsymbol{\beta}_{10} + \boldsymbol{\epsilon}_1$, where $\boldsymbol{\epsilon}_1 = \boldsymbol{\beta}_1 - \boldsymbol{\beta}_{10} = (\epsilon_{1(1)}, \ldots, \epsilon_{1(p_1)})^\top$ with $\epsilon_{1(r)} = O(n^{-1/2})$ for $r = 1, \ldots, p_1$. Now, some notation is in order. Let

$$\boldsymbol{\epsilon}_1^* = \begin{bmatrix} \boldsymbol{I}_{p_1} \\ -(\boldsymbol{X}_2^\top \boldsymbol{\Phi}\boldsymbol{W}\boldsymbol{X}_2)^{-1}\boldsymbol{X}_2^\top \boldsymbol{\Phi}\boldsymbol{W}\boldsymbol{X}_1 \end{bmatrix} \boldsymbol{\epsilon}_1.$$

We define the $n \times n$ matrices $\boldsymbol{Z}_\beta = \boldsymbol{X}(\boldsymbol{X}^\top \boldsymbol{\Phi}\boldsymbol{W}\boldsymbol{X})^{-1}\boldsymbol{X}^\top$ and $\boldsymbol{Z}_{\beta_2} = \boldsymbol{X}_2(\boldsymbol{X}_2^\top \boldsymbol{\Phi}\boldsymbol{W}\boldsymbol{X}_2)^{-1}\boldsymbol{X}_2^\top$. Also, let $\boldsymbol{F} = \mathrm{diag}\{f_1, \ldots, f_n\}$, $\boldsymbol{G} = \mathrm{diag}\{g_1, \ldots, g_n\}$, $\boldsymbol{t}_1 = (t_{11}, \ldots, t_{1n})^\top = \boldsymbol{X}\boldsymbol{\epsilon}_1^*$, $\boldsymbol{e}_1 = (e_{11}, \ldots, e_{1n})^\top = \boldsymbol{X}_1\boldsymbol{\epsilon}_1$, $\boldsymbol{T}_1 = \mathrm{diag}\{t_{11}, \ldots, t_{1n}\}$, $\boldsymbol{T}_1^{(2)} = \boldsymbol{T}_1 \odot \boldsymbol{T}_1$, $\boldsymbol{T}_1^{(3)} = \boldsymbol{T}_1^{(2)} \odot \boldsymbol{T}_1$, $\boldsymbol{E}_1 = \mathrm{diag}\{e_{11}, \ldots, e_{1n}\}$, and

$$f_l = \frac{1}{V_l}\frac{d\mu_l}{d\eta_l}\frac{d^2\mu_l}{d\eta_l^2}, \quad g_l = f_l - \frac{1}{V_l^2}\frac{dV_l}{d\mu_l}\left(\frac{d\mu_l}{d\eta_l}\right)^3, \quad l = 1, \ldots, n.$$

The nonnull cumulative distribution function of S_{T} under Pitman alternatives for testing $\mathcal{H}_0^1 : \boldsymbol{\beta}_1 = \boldsymbol{\beta}_{10}$ in GLMs with dispersion covariates can be expressed as

$$\Pr(S_T' \leq x|\mathcal{H}_{an}^1) = G_{p_1,\lambda'}(x) + \sum_{k=0}^{3} b_k' G_{p_1+2k,\lambda'}(x) + O(n^{-1}),$$

where $\lambda' = (1/2)\text{tr}\{\boldsymbol{K}_{\beta 11.2}\boldsymbol{\epsilon}_1\boldsymbol{\epsilon}_1^{\top}\}$, and

$$\boldsymbol{K}_{\beta 11.2} = \boldsymbol{X}_1^{\top}\boldsymbol{\Phi}\boldsymbol{W}\boldsymbol{X}_1 - \boldsymbol{X}_1^{\top}\boldsymbol{\Phi}\boldsymbol{W}\boldsymbol{X}_2(\boldsymbol{X}_2^{\top}\boldsymbol{\Phi}\boldsymbol{W}\boldsymbol{X}_2)^{-1}\boldsymbol{X}_2^{\top}\boldsymbol{\Phi}\boldsymbol{W}\boldsymbol{X}_1.$$

The coefficients b_k's can be written in matrix notation as

$$\begin{aligned} b_1' &= -\frac{1}{4}\text{tr}\{\boldsymbol{\Phi}[(3\boldsymbol{F} + 2\boldsymbol{G})\boldsymbol{Z}_{\beta_2 d}\boldsymbol{T}_1 - (\boldsymbol{F} + 2\boldsymbol{G})\boldsymbol{Z}_{\beta d}\boldsymbol{T}_1]\} \\ &\quad + \frac{1}{2}\text{tr}\{\boldsymbol{\Phi}[(\boldsymbol{F} + \boldsymbol{G})\boldsymbol{E}_1\boldsymbol{T}_1^{(2)} - \boldsymbol{F}\boldsymbol{T}_1^{(3)}]\}, \\ b_2' &= \frac{1}{4}\text{tr}\{\boldsymbol{\Phi}\boldsymbol{F}\boldsymbol{T}_1^{(3)}\} - \frac{1}{4}\text{tr}\{\boldsymbol{\Phi}(\boldsymbol{F} + 2\boldsymbol{G})(\boldsymbol{Z}_{\beta} - \boldsymbol{Z}_{\beta_2})_d\boldsymbol{T}_1\}, \\ b_3' &= -\frac{1}{12}\text{tr}\{\boldsymbol{\Phi}(\boldsymbol{F} + 2\boldsymbol{G})\boldsymbol{T}_1^{(3)}\}, \end{aligned}$$

and $b_0' = -(b_1' + b_2' + b_3')$. The subscript d indicates that the off-diagonal elements of the matrix were set equal to zero. The b_k's are of order $O(n^{-1/2})$ and all quantities, except $\boldsymbol{\epsilon}_1$, are evaluated under the null hypothesis $\mathcal{H}_0^1$.

A brief commentary on these coefficients is in order. It is interesting to note that the b_k's are functions of the model matrix $\boldsymbol{X}$ and of the diagonal matrix $\boldsymbol{\Phi}$. These coefficients depend on the mean link function and its first and second derivatives. They also involve the variance function and its first derivative. Note that these coefficients involve the inverse of the dispersion link function. Unfortunately, they are not easy to interpret in generality and provide no indication as to what structural aspects of the model contribute significantly to their magnitude.

Some simplifications in the coefficients b_k's can be achieved by examining special cases. For example, for any distribution in the exponential family Eq. (2.9) with identity mean link function ($d(\mu_l) = \mu_l$), which implies that $f_l = 0$ and $g_l = -V_l^{-2}dV_l/d\mu_l$ ($l = 1, \ldots, n$), we have

$$\begin{aligned} b_1' &= \frac{1}{2}\text{tr}\{\boldsymbol{\Phi}\boldsymbol{G}\boldsymbol{E}_1\boldsymbol{T}_1^{(2)}\} + \frac{1}{2}\text{tr}\{\boldsymbol{\Phi}\boldsymbol{G}(\boldsymbol{Z}_{\beta} - \boldsymbol{Z}_{\beta_2})_d\boldsymbol{T}_1\}, \\ b_2' &= -\frac{1}{2}\text{tr}\{\boldsymbol{\Phi}\boldsymbol{G}(\boldsymbol{Z}_{\beta} - \boldsymbol{Z}_{\beta_2})_d\boldsymbol{T}_1\}, \quad b_3' = -\frac{1}{6}\text{tr}\{\boldsymbol{\Phi}\boldsymbol{G}\boldsymbol{T}_1^{(3)}\}, \end{aligned}$$

and $b_0' = -(b_1' + b_2' + b_{i3}')$. As expected, the above coefficients vanish for the normal model since the nonnull distribution function of the gradient criteria agrees exactly with the $\chi^2_{p_1,\lambda'}$ distribution. For the simple model

$d(\mu_l) = \mu_l = \mu$ and for testing $\mu = \mu_0$, where μ_0 is a specified value, we have that

$$b'_1 = -\frac{(\mu-\mu_0)^3}{2V^2}\frac{dV}{d\mu}\sum_{l=1}^{n}\phi_l - \frac{(\mu-\mu_0)}{2}\frac{dV}{d\mu},$$

$$b'_2 = \frac{(\mu-\mu_0)}{2}\frac{dV}{d\mu}, \quad b'_3 = \frac{(\mu-\mu_0)^3}{6V^2}\frac{dV}{d\mu}\sum_{l=1}^{n}\phi_l,$$

and $b'_0 = -(b'_1 + b'_2 + b'_{i3})$, where is assumed that $\mu - \mu_0$ is of order $O(n^{-1/2})$.

2.7.2 Nonnull Distribution Under $\mathcal{H}^2_{\mathrm{an}} : \boldsymbol{\gamma}_1 = \boldsymbol{\gamma}_{10} + \boldsymbol{\epsilon}_2$

We assume the local alternative hypothesis $\mathcal{H}^2_{\mathrm{an}} : \boldsymbol{\gamma}_1 = \boldsymbol{\gamma}_{10} + \boldsymbol{\epsilon}_2$, where $\boldsymbol{\epsilon}_2 = \boldsymbol{\gamma}_1 - \boldsymbol{\gamma}_{10} = (\epsilon_{2(1)}, \ldots, \epsilon_{2(q_1)})^\top$ with $\epsilon_{2(r)} = O(n^{-1/2})$ for $r = 1, \ldots, q_1$. We present the nonnull asymptotic expansion up to order $O(n^{-1/2})$ for the nonnull cumulative distribution function of the gradient statistic for testing a subset of dispersion parameters in GLMs with dispersion covariates. We introduce the following quantities:

$$\boldsymbol{\epsilon}_2^* = \begin{bmatrix} \boldsymbol{I}_{q_1} \\ -(\boldsymbol{S}_2^\top(-\boldsymbol{D}_2\boldsymbol{\Phi}_1^{(2)})\boldsymbol{S}_2)^{-1}\boldsymbol{S}_2^\top(-\boldsymbol{D}_2\boldsymbol{\Phi}_1^{(2)})\boldsymbol{S}_1 \end{bmatrix}\boldsymbol{\epsilon}_2,$$

$\boldsymbol{Z}_\gamma = \boldsymbol{S}(-\boldsymbol{S}^\top\boldsymbol{D}_2\boldsymbol{\Phi}_1^{(2)}\boldsymbol{S})^{-1}\boldsymbol{S}^\top$, $\boldsymbol{Z}_{\gamma_2} = \boldsymbol{S}_2(-\boldsymbol{S}_2^\top\boldsymbol{D}_2\boldsymbol{\Phi}_1^{(2)}\boldsymbol{S}_2)^{-1}\boldsymbol{S}_2^\top$, $\boldsymbol{t}_2 = (t_{21}, \ldots, t_{2n})^\top = \boldsymbol{S}\boldsymbol{\epsilon}_2^*$, $\boldsymbol{e}_2 = (e_{21}, \ldots, e_{2n})^\top = \boldsymbol{S}_1\boldsymbol{\epsilon}_2$, $\boldsymbol{E}_2 = \mathrm{diag}\{e_{21}, \ldots, e_{2n}\}$, $\boldsymbol{T}_2 = \mathrm{diag}\{t_{21}, \ldots, t_{2n}\}$, $\boldsymbol{T}_2^{(2)} = \boldsymbol{T}_2 \odot \boldsymbol{T}_2$, $\boldsymbol{T}_2^{(3)} = \boldsymbol{T}_2^{(2)} \odot \boldsymbol{T}_2$, $\boldsymbol{\Phi}_2 = \mathrm{diag}\{\phi_{21}, \ldots, \phi_{2n}\}$, and $\boldsymbol{D}_3 = \mathrm{diag}\{c_{31}, \ldots, c_{3n}\}$.

The nonnull cumulative distribution function of S''_{T} under Pitman alternatives for testing $\mathcal{H}^2_0 : \boldsymbol{\gamma}_1 = \boldsymbol{\gamma}_{10}$ in GLMs with dispersion covariates can be expressed as

$$\Pr(S''_{\mathrm{T}} \le x|\mathcal{H}^2_{\mathrm{an}}) = G_{q_1,\lambda''}(x) + \sum_{k=0}^{3} b''_k G_{q_1+2k,\lambda''}(x) + O(n^{-1}),$$

where $\lambda'' = (1/2)\mathrm{tr}\{\boldsymbol{K}_{\gamma 11.2}\boldsymbol{\epsilon}_2\boldsymbol{\epsilon}_2^\top\}$, and

$$\boldsymbol{K}_{\gamma 11.2} = -\boldsymbol{S}_1^\top(-\boldsymbol{D}_2\boldsymbol{\Phi}_1^{(2)})\boldsymbol{S}_2(\boldsymbol{S}_2^\top(-\boldsymbol{D}_2\boldsymbol{\Phi}_1^{(2)})\boldsymbol{S}_2)^{-1}\boldsymbol{S}_2^\top(-\boldsymbol{D}_2\boldsymbol{\Phi}_1^{(2)})\boldsymbol{S}_1 + \boldsymbol{S}_1^\top(-\boldsymbol{D}_2\boldsymbol{\Phi}_1^{(2)})\boldsymbol{S}_1,$$

and, after some algebra, the coefficients b''_ks take the forms

$$b''_1 = -\frac{1}{4}\mathrm{tr}\{(\boldsymbol{D}_3\boldsymbol{\Phi}_1^{(3)} + 3\boldsymbol{D}_2\boldsymbol{\Phi}_1\boldsymbol{\Phi}_2)\boldsymbol{Z}_{\gamma d}\boldsymbol{T}_2\}$$

$$+\frac{1}{2}\mathrm{tr}\{(\boldsymbol{D}_3\boldsymbol{\Phi}_1^{(3)}+\boldsymbol{D}_2\boldsymbol{\Phi}_1\boldsymbol{\Phi}_2)\boldsymbol{T}_2^{(3)}\}+\frac{1}{2}\mathrm{tr}\{\boldsymbol{\Phi}_1\boldsymbol{W}\boldsymbol{T}_2\boldsymbol{Z}_{\beta d}\}$$
$$+\frac{1}{4}\mathrm{tr}\{(3\boldsymbol{D}_3\boldsymbol{\Phi}_1^{(3)}+5\boldsymbol{D}_2\boldsymbol{\Phi}_1\boldsymbol{\Phi}_2)\boldsymbol{Z}_{\gamma_2 d}\boldsymbol{T}_2\}$$
$$-\frac{1}{2}\mathrm{tr}\{(\boldsymbol{D}_3\boldsymbol{\Phi}_1^{(3)}+2\boldsymbol{D}_2\boldsymbol{\Phi}_1\boldsymbol{\Phi}_2)\boldsymbol{E}_2\boldsymbol{T}_2^{(2)}\},$$
$$b_2''=\frac{1}{4}\mathrm{tr}\{(\boldsymbol{D}_3\boldsymbol{\Phi}_1^{(3)}+3\boldsymbol{D}_2\boldsymbol{\Phi}_1\boldsymbol{\Phi}_2)(\boldsymbol{Z}_\gamma-\boldsymbol{Z}_{\gamma_2})_d\boldsymbol{T}_2\}$$
$$-\frac{1}{4}\mathrm{tr}\{(\boldsymbol{D}_3\boldsymbol{\Phi}_1^{(3)}+\boldsymbol{D}_2\boldsymbol{\Phi}_1\boldsymbol{\Phi}_2)\boldsymbol{T}_2^{(3)}\},$$
$$b_3''=\frac{1}{12}\mathrm{tr}\{(\boldsymbol{D}_3\boldsymbol{\Phi}_1^{(3)}+3\boldsymbol{D}_2\boldsymbol{\Phi}_1\boldsymbol{\Phi}_2)\boldsymbol{T}_2^{(3)}\},$$

and $b_0''=-(b_1''+b_2''+b_3'')$. The b_k''s are of order $O(n^{-1/2})$ and all quantities, except $\boldsymbol{\epsilon}_2$, are evaluated under the null hypothesis $\mathcal{H}_0^2$. It is interesting to note that the b_k''s are functions of the model matrix $\boldsymbol{X}$ and of the matrix $\boldsymbol{S}$. They involve the dispersion link function and its first and second derivatives. They also depend on the mean link function through its first derivative and they involve the variance function. Regrettably, these equations are very difficult to interpret.

For a GLM with dispersion covariates with identity dispersion link function ($h(\phi_l)=\phi_l$), the coefficients which define the nonnull cumulative distribution function of the gradient test statistic for testing the null hypothesis $\mathcal{H}_0^2:\boldsymbol{\gamma}_1=\boldsymbol{\gamma}_{10}$ can be written as

$$b_1''=\frac{1}{4}\mathrm{tr}\{\boldsymbol{D}_3(-\boldsymbol{Z}_{\gamma d}\boldsymbol{T}_2+2\boldsymbol{T}_2^{(3)}+3\boldsymbol{Z}_{\gamma_2 d}\boldsymbol{T}_2-2\boldsymbol{E}_2\boldsymbol{T}_2^{(2)})\}$$
$$+\frac{1}{2}\mathrm{tr}\{\boldsymbol{W}\boldsymbol{T}_2\boldsymbol{Z}_{\beta d}\},$$
$$b_2''=\frac{1}{4}\mathrm{tr}\{\boldsymbol{D}_3(\boldsymbol{Z}_\gamma-\boldsymbol{Z}_{\gamma_2})_d\boldsymbol{T}_2\}-\frac{1}{4}\mathrm{tr}\{\boldsymbol{D}_3\boldsymbol{T}_2^{(3)}\},\quad b_3''=\frac{1}{12}\mathrm{tr}\{\boldsymbol{D}_3\boldsymbol{T}_2^{(3)}\},$$

and $b_0''=-(b_1''+b_2''+b_3'')$. By considering the dispersion link function $h(\phi_l)=\log\phi_l$ for the normal and inverse Gaussian models, the b_k''s reduce to

$$b_1''=\frac{1}{8}\mathrm{tr}\{2\boldsymbol{T}_2^{(3)}+(\boldsymbol{Z}_\gamma+\boldsymbol{Z}_{\gamma_2})_d\boldsymbol{T}_2\}+\frac{1}{2}\mathrm{tr}\{\boldsymbol{\Phi}_1\boldsymbol{W}\boldsymbol{T}_2\boldsymbol{Z}_{\beta d}\},$$
$$b_2''=-\frac{1}{8}\mathrm{tr}\{(\boldsymbol{Z}_\gamma-\boldsymbol{Z}_{\gamma_2})_d\boldsymbol{T}_2+\boldsymbol{T}_2^{(3)}\},\quad b_3''=-\frac{1}{24}\mathrm{tr}\{\boldsymbol{T}_2^{(3)}\},$$

and $b_0''=-(b_1''+b_2''+b_3'')$.

If $h(\phi_l) = \phi_l = \phi$ for all $l = 1, \ldots, n$, then the nonnull cumulative distribution function of the gradient statistic for testing $\mathcal{H}_0^2 : \phi = \phi_0$ under the local alternative $\mathcal{H}_{\text{an}}^2 : \phi = \phi_0 + n^{-1/2}\epsilon_2$ is given by Eq. (2.2) with $q = 1$, noncentrality parameter $\lambda'' = -\ddot{c}(\phi)\epsilon_2^2/2$,

$$b_1'' = \frac{p\epsilon_2}{2\phi} + \frac{\dddot{c}(\phi)\epsilon_2}{4\ddot{c}(\phi)}, \quad b_2'' = -\frac{\dddot{c}(\phi)\epsilon_2}{4\ddot{c}(\phi)} - \frac{\dddot{c}(\phi)\epsilon_2^3}{4}, \quad b_3'' = \frac{\dddot{c}(\phi)\epsilon_2^3}{12},$$

and $b_0'' = -(b_1'' + b_2'' + b_3'')$, where $\epsilon_2 = \sqrt{n}(\phi - \phi_0)$ and $\dddot{c}(\phi) = d\ddot{c}(\phi)/d\phi$. It should be noticed that the above expressions depend on the model only through ϕ and the rank of the model matrix $\boldsymbol{X}$. They do not involve the unknown parameter vector $\boldsymbol{\beta}$.

2.7.3 Power Comparisons

First, we shall compare the local powers of the rival tests (ie, the LR, Wald, score, and gradient tests) for testing the null hypothesis $\mathcal{H}_0^1 : \boldsymbol{\beta}_1 = \boldsymbol{\beta}_{10}$ in the class of GLMs with dispersion covariates. Let Π_i' be the power function, up to order $O(n^{-1/2})$, of the test that uses the statistic S_i', for $i = \text{LR, W, R, T}$. We have

$$\Pi_i' - \Pi_j' = \sum_{k=0}^{3}(b_{jk}' - b_{ik}')G_{p_1+2k,\lambda'}(x), \tag{2.11}$$

for $i \neq j$, and x is replaced by the appropriate quantile of the reference distribution for the chosen nominal level. The coefficients that define the local power functions of the LR, Wald, and score tests can be found in Lemonte [18]. After some algebra, from Eqs. (2.11) and (2.7) we have

$$\begin{aligned}
\Pi_{\text{LR}}' - \Pi_{\text{T}}' &= k_1' g_{p-p_1+4,\lambda'}(x) + k_2' g_{p-p_1+6,\lambda'}(x),\\
\Pi_{\text{W}}' - \Pi_{\text{T}}' &= k_3' g_{p-p_1+4,\lambda'}(x) + k_4' g_{p-p_1+6,\lambda'}(x),\\
\Pi_{\text{R}}' - \Pi_{\text{T}}' &= k_5' g_{p-p_1+4,\lambda'}(x) + k_6' g_{p-p_1+6,\lambda'}(x),\\
\Pi_{\text{LR}}' - \Pi_{\text{W}}' &= k_7' g_{p-p_1+4,\lambda'}(x) + k_8' g_{p-p_1+6,\lambda'}(x),\\
\Pi_{\text{LR}}' - \Pi_{\text{R}}' &= k_9' g_{p-p_1+4,\lambda'}(x) + k_{10}' g_{p-p_1+6,\lambda'}(x),\\
\Pi_{\text{W}}' - \Pi_{\text{R}}' &= k_{11}' g_{p-p_1+4,\lambda'}(x) + k_{12}' g_{p-p_1+6,\lambda'}(x),
\end{aligned} \tag{2.12}$$

where

$$k_1' = \frac{1}{2}\text{tr}\{\boldsymbol{\Phi}(\boldsymbol{F} + 2\boldsymbol{G})(\boldsymbol{Z}_{\beta} - \boldsymbol{Z}_{\beta_2})_d \boldsymbol{T}_1\},$$

$$k_2' = \frac{1}{6}\text{tr}\{\boldsymbol{\Phi}(\boldsymbol{F} + 2\boldsymbol{G})\boldsymbol{T}_1^{(3)}\}, \quad k_3' = 3k_1', \quad k_4' = 3k_2',$$

$$k_5' = k_1' + \text{tr}\{\boldsymbol{\Phi}(\boldsymbol{F} - \boldsymbol{G})(\boldsymbol{Z}_{\beta} - \boldsymbol{Z}_{\beta_2})_d \boldsymbol{T}_1\},$$

$$k'_6 = \frac{1}{2}\mathrm{tr}\{\boldsymbol{\Phi} \boldsymbol{F} \boldsymbol{T}_1^{(3)}\}, \quad k'_7 = -2k'_1,$$

$$k'_8 = -2k'_2, \quad k'_9 = k'_1 - k'_5, \quad k'_{10} = -\frac{1}{3}\mathrm{tr}\{\boldsymbol{\Phi}(\boldsymbol{F} - \boldsymbol{G})\boldsymbol{T}_1^{(3)}\},$$

$$k'_{11} = 3\mathrm{tr}\{\boldsymbol{\Phi} \boldsymbol{G}(\boldsymbol{Z}_\beta - \boldsymbol{Z}_{\beta_2})_d \boldsymbol{T}_1\}, \quad k'_{12} = \mathrm{tr}\{\boldsymbol{\Phi} \boldsymbol{G} \boldsymbol{T}_1^{(3)}\}.$$

We arrive at the following general conclusions from Eq. (2.12). For GLMs with dispersion covariates with canonical mean link function ($\boldsymbol{G} = \boldsymbol{0}_{n,n}$), we have $k'_{11} = k'_{12} = 0$ and hence $\Pi'_{\mathrm{W}} = \Pi'_{\mathrm{R}}$. However, if $k'_{11} \geq 0$ and $k'_{12} \geq 0$ with $k'_{11} + k'_{12} > 0$ we have $\Pi'_{\mathrm{W}} > \Pi'_{\mathrm{R}}$, and if $k'_{11} \leq 0$ and $k'_{12} \leq 0$ with $k'_{11} + k'_{12} < 0$ we have $\Pi'_{\mathrm{W}} < \Pi'_{\mathrm{R}}$. It is possible to show that $\Pi'_{\mathrm{LR}} = \Pi'_{\mathrm{W}} = \Pi'_{\mathrm{T}}$ if $\boldsymbol{F} = -2\boldsymbol{G}$, that is,

$$\frac{d^2\mu_l}{d\eta_l^2} = \frac{2}{3V_l}\frac{dV_l}{d\mu_l}\left(\frac{d\mu_l}{d\eta_l}\right)^2, \quad l = 1, \ldots, n.$$

The GLMs with dispersion covariates for which this equality holds have the mean link function defined by

$$\eta_l = \int V_l^{-3/2}\, d\mu_l, \quad l = 1, \ldots, n.$$

For the gamma model this function is $\eta_l = \mu_l^{-1/3}$ ($l = 1, \ldots, n$). Additionally, we have that $\Pi'_{\mathrm{R}} = \Pi'_{\mathrm{T}}$ for any GLM with dispersion covariates with identity mean link function, that is, $\boldsymbol{F} = \boldsymbol{0}_{n,n}$. If $k'_5 \geq 0$ and $k'_6 \geq 0$ with $k'_5 + k'_6 > 0$ we have $\Pi'_{\mathrm{R}} > \Pi'_{\mathrm{T}}$, and if $k'_5 \leq 0$ and $k'_6 \leq 0$ with $k'_5 + k'_6 < 0$ we have $\Pi'_{\mathrm{R}} < \Pi'_{\mathrm{T}}$. Also, $\Pi'_{\mathrm{LR}} = \Pi'_{\mathrm{R}}$ if $k_9 = k_{10} = 0$, that is, $\boldsymbol{F} = \boldsymbol{G}$, which occurs only for normal models with any mean link function. Now, if $k'_9 \geq 0$ and $k'_{10} \geq 0$ with $k'_9 + k'_{10} > 0$ we have $\Pi'_{\mathrm{LR}} > \Pi'_{\mathrm{R}}$, and if $k'_9 \leq 0$ and $k'_{10} \leq 0$ with $k'_9 + k'_{10} < 0$ we have $\Pi'_{\mathrm{LR}} < \Pi'_{\mathrm{R}}$. Finally, the equality $\Pi'_{\mathrm{LR}} = \Pi'_{\mathrm{W}} = \Pi'_{\mathrm{R}} = \Pi'_{\mathrm{T}}$ holds only for normal models with identity mean link function.

Next, we shall compare the local powers of the rival tests for testing the null hypothesis $\mathcal{H}_0^2 : \boldsymbol{\gamma}_1 = \boldsymbol{\gamma}_{10}$ in the class of GLMs with dispersion covariates. Let Π''_i be the power function, up to order $O(n^{-1/2})$, of the test that uses the statistic S''_i, for $i = \mathrm{LR}, \mathrm{W}, \mathrm{R}, \mathrm{T}$. We can show, after some algebra, that

$$\begin{aligned}
\Pi''_{\rm LR} - \Pi''_{\rm W} &= k''_1 g_{q-q_1+4,\lambda''}(x) + k''_2 g_{q-q_1+6,\lambda''}(x),\\
\Pi''_{\rm LR} - \Pi''_{\rm R} &= k''_3 g_{q-q_1+4,\lambda''}(x) + k''_4 g_{q-q_1+6,\lambda''}(x),\\
\Pi''_{\rm LR} - \Pi''_{\rm T} &= k''_5 g_{q-q_1+4,\lambda''}(x) + k''_6 g_{q-q_1+6,\lambda''}(x),\\
\Pi''_{\rm W} - \Pi''_{\rm R} &= k''_7 g_{q-q_1+4,\lambda''}(x) + k''_8 g_{q-q_1+6,\lambda''}(x),\\
\Pi''_{\rm W} - \Pi''_{\rm T} &= k''_9 g_{q-q_1+4,\lambda''}(x) + k''_{10} g_{q-q_1+6,\lambda''}(x),\\
\Pi''_{\rm R} - \Pi''_{\rm T} &= k''_{11} g_{q-q_1+4,\lambda''}(x) + k''_{12} g_{q-q_1+6,\lambda''}(x),
\end{aligned} \tag{2.13}$$

where

$$\begin{aligned}
k''_1 &= \mathrm{tr}\{(\boldsymbol{D}_3\boldsymbol{\Phi}_1^{(3)} + 3\boldsymbol{D}_2\boldsymbol{\Phi}_1\boldsymbol{\Phi}_2)(\boldsymbol{Z}_\gamma - \boldsymbol{Z}_{\gamma_2})_d\boldsymbol{T}_2\},\\
k''_2 &= \frac{1}{3}\mathrm{tr}\{(\boldsymbol{D}_3\boldsymbol{\Phi}_1^{(3)} + 3\boldsymbol{D}_2\boldsymbol{\Phi}_1\boldsymbol{\Phi}_2)\boldsymbol{T}_2^{(3)}\},\\
k''_3 &= \mathrm{tr}\{\boldsymbol{D}_3\boldsymbol{\Phi}_1^{(3)}(\boldsymbol{Z}_\gamma - \boldsymbol{Z}_{\gamma_2})_d\boldsymbol{T}_2\}, \quad k''_4 = \frac{1}{3}\mathrm{tr}\{\boldsymbol{D}_3\boldsymbol{\Phi}_1^{(3)}\boldsymbol{T}_2^{(3)}\},\\
k''_5 &= -\frac{1}{2}k''_1, \quad k''_6 = -\frac{1}{2}k''_2, \quad k''_9 = -\frac{3}{2}k''_1, \quad k''_{10} = -\frac{3}{2}k''_2,\\
k''_7 &= -3\mathrm{tr}\{\boldsymbol{D}_2\boldsymbol{\Phi}_1\boldsymbol{\Phi}_2(\boldsymbol{Z}_\gamma - \boldsymbol{Z}_{\gamma_2})_d\boldsymbol{T}_2\}, \quad k''_8 = -\mathrm{tr}\{\boldsymbol{D}_2\boldsymbol{\Phi}_1\boldsymbol{\Phi}_2\boldsymbol{T}_2^{(3)}\},\\
k''_{11} &= -\frac{3}{2}\mathrm{tr}\{(\boldsymbol{D}_3\boldsymbol{\Phi}_1^{(3)} + \boldsymbol{D}_2\boldsymbol{\Phi}_1\boldsymbol{\Phi}_2)(\boldsymbol{Z}_\gamma - \boldsymbol{Z}_{\gamma_2})_d\boldsymbol{T}_2\},\\
k''_{12} &= -\frac{1}{2}\mathrm{tr}\{(\boldsymbol{D}_3\boldsymbol{\Phi}_1^{(3)} + \boldsymbol{D}_2\boldsymbol{\Phi}_1\boldsymbol{\Phi}_2)\boldsymbol{T}_2^{(3)}\}.
\end{aligned}$$

From Eq. (2.13) we have $\Pi''_{\rm W} > \Pi''_{\rm R}$ if $k''_7 \geq 0$ and $k''_8 \geq 0$ with $k''_7 + k''_8 > 0$, and if $k''_7 \leq 0$ and $k''_8 \leq 0$ with $k''_7 + k''_8 < 0$ we have $\Pi''_{\rm W} < \Pi''_{\rm R}$. Also, $\Pi''_{\rm W} = \Pi''_{\rm R}$ if $k''_7 = k''_8 = 0$, that is, $\boldsymbol{\Phi}_2 = \boldsymbol{0}_{n,n}$, which occurs only for GLMs with dispersion covariates with identity dispersion link function. Additionally, Eq. (2.13) reveals that with the exception of the Wald and score tests, it is not possible to have any other equality among the power functions in the class of GLMs with dispersion covariates for testing the null hypothesis $\mathcal{H}_0^2 : \boldsymbol{\gamma}_1 = \boldsymbol{\gamma}_{10}$. It implies that only strict inequality holds for any other power comparison among the power functions of the tests that are based on the statistics $S''_{\rm LR}$, $S''_{\rm W}$, $S''_{\rm R}$, and $S''_{\rm T}$. For example, from Eq. (2.13) we have $\Pi''_{\rm LR} > \Pi''_{\rm W}$ ($\Pi''_{\rm LR} < \Pi''_{\rm W}$) if $k''_1 \geq 0$ and $k''_2 \geq 0$ with $k''_1 + k''_2 > 0$ (if $k''_1 \leq 0$ and $k''_2 \leq 0$ with $k''_1 + k''_2 < 0$), and so on.

Finally, we present an analytical comparison among the local powers of the four tests for testing the null hypothesis $\mathcal{H}_0^2 : \phi = \phi_0$. We have

$$\Pi''_{\rm LR} - \Pi''_{\rm W} = \Pi''_{\rm LR} - \Pi''_{\rm R} = -\frac{\dddot{c}(\phi_0)\epsilon}{\ddot{c}(\phi_0)\sqrt{n}} g_{5,\lambda''}(x) + \frac{\dddot{c}(\phi_0)\epsilon^3}{6\sqrt{n}} g_{7,\lambda''}(x),$$

$$\Pi''_{\mathrm{LR}} - \Pi''_{\mathrm{T}} = \frac{\dddot{c}(\phi_0)\epsilon}{2\ddot{c}(\phi_0)\sqrt{n}} g_{5,\lambda''}(x) - \frac{\dddot{c}(\phi_0)\epsilon^3}{6\sqrt{n}} g_{7,\lambda''}(x), \quad \Pi''_{\mathrm{W}} - \Pi''_{\mathrm{R}} = 0,$$

$$\Pi''_{\mathrm{W}} - \Pi''_{\mathrm{T}} = \Pi_{\mathrm{R}} - \Pi_{\mathrm{T}} = \frac{3\dddot{c}(\phi_0)\epsilon}{2\ddot{c}(\phi_0)\sqrt{n}} g_{5,\lambda''}(x) - \frac{\dddot{c}(\phi_0)\epsilon^3}{2\sqrt{n}} g_{7,\lambda''}(x).$$

For example, we have $c(\phi) = (1/2)\log\phi$ for the normal and inverse Gaussian models, which implies that $\dot{c}(\phi) = 1/(2\phi)$, $\ddot{c}(\phi) = -1/(2\phi^2)$, and $\dddot{c}(\phi) = 1/\phi^3$. Hence, from the above expressions we arrive at the following inequalities: $\Pi''_{\mathrm{T}} > \Pi''_{\mathrm{LR}} > \Pi''_{\mathrm{W}} = \Pi''_{\mathrm{R}}$ if $\phi > \phi_0$, and $\Pi''_{\mathrm{T}} < \Pi''_{\mathrm{LR}} < \Pi''_{\mathrm{W}} = \Pi''_{\mathrm{R}}$ if $\phi < \phi_0$.

We have from the power comparisons presented in this section that there is no uniform superiority of one test with respect to the others for testing the null hypotheses $\mathcal{H}_0^1 : \boldsymbol{\beta}_1 = \boldsymbol{\beta}_{10}$ and $\mathcal{H}_0^2 : \boldsymbol{\gamma}_1 = \boldsymbol{\gamma}_{10}$ in the class of GLMs with dispersion covariates. Therefore, the gradient statistic, which is very simple to be computed, can be an interesting alternative to the other statistics for testing hypotheses in this class of models.

2.8 CENSORED EXPONENTIAL REGRESSION MODEL

In Section 1.6 we consider the class of exponential regression models (ExpRMs), which is defined in Eq. (1.5). The Monte Carlo simulation experiments revealed that the gradient test that uses the statistic in Eq. (1.8) is very competitive to the LR test for testing inference in this class of regression models in terms of finite-sample size properties. We present in this section the asymptotic expansion up to order $O(n^{-1/2})$ for the nonnull cumulative distribution function of the gradient statistic for testing a subset of regression parameters in ExpRMs and provide local power comparisons between the gradient and LR tests. In what follows, we use the same notation as that of Section 1.6.

We shall assume that the local alternative hypothesis is $\mathcal{H}_{\mathrm{an}} : \boldsymbol{\beta}_1 = \boldsymbol{\beta}_{10} + \boldsymbol{\epsilon}$, where $\boldsymbol{\epsilon} = \boldsymbol{\beta}_1 - \boldsymbol{\beta}_{10} = (\epsilon_1, \ldots, \epsilon_q)^\top$ with $\epsilon_s = O(n^{-1/2})$ for $s = 1, \ldots, q$. We introduce the following quantities:

$$\boldsymbol{Z} = \boldsymbol{X}(\boldsymbol{X}^\top \boldsymbol{W}\boldsymbol{X})^{-1}\boldsymbol{X}^\top = ((z_{lm})),$$

$$\boldsymbol{Z}_2 = \boldsymbol{X}_2(\boldsymbol{X}_2^\top \boldsymbol{W}\boldsymbol{X}_2)^{-1}\boldsymbol{X}_2^\top = ((z_{2lm})),$$

$$\boldsymbol{Z}_d = \mathrm{diag}\{z_{11}, \ldots, z_{nn}\}, \quad \boldsymbol{Z}_{2d} = \mathrm{diag}\{z_{211}, \ldots, z_{2nn}\},$$

$$\boldsymbol{W}' = \mathrm{diag}\{w'_1, \ldots, w'_n\}, \quad \boldsymbol{\epsilon}^* = \begin{bmatrix} \boldsymbol{I}_q \\ -(\boldsymbol{X}_2^\top \boldsymbol{W}\boldsymbol{X}_2)^{-1}\boldsymbol{X}_2^\top \boldsymbol{W}\boldsymbol{X}_1 \end{bmatrix} \boldsymbol{\epsilon},$$

where $w'_l = dw_l/d\mu_l$, for $l = 1, \ldots, n$. Also, let $\boldsymbol{b} = (b_1, \ldots, b_n)^\top = \boldsymbol{X}\boldsymbol{\epsilon}^*$, $\boldsymbol{e} = (e_1, \ldots, e_n)^\top = \boldsymbol{X}_1\boldsymbol{\epsilon}$, $\boldsymbol{B} = \text{diag}\{b_1, \ldots, b_n\}$, $\boldsymbol{E} = \text{diag}\{e_1, \ldots, e_n\}$, $\boldsymbol{B}^{(2)} = \boldsymbol{B} \odot \boldsymbol{B}$ and $\boldsymbol{B}^{(3)} = \boldsymbol{B}^{(2)} \odot \boldsymbol{B}$.

For testing the null hypothesis $\mathcal{H}_0 : \boldsymbol{\beta}_1 = \boldsymbol{\beta}_{10}$ in the class of censored ExpRMs, the nonnull cumulative distribution function of S_T under Pitman alternatives can be expressed as

$$\Pr(S_\mathrm{T} \leq x|\mathcal{H}_{\text{an}}) = G_{q,\lambda}(x) + \sum_{k=0}^{3} b_k G_{q+2k,\lambda}(x) + O(n^{-1}),$$

where $\lambda = (1/2)\boldsymbol{\epsilon}^\top[\boldsymbol{X}_1^\top \boldsymbol{W}\boldsymbol{X}_1 - \boldsymbol{X}_1^\top \boldsymbol{W}\boldsymbol{X}_2(\boldsymbol{X}_2^\top \boldsymbol{W}\boldsymbol{X}_2)^{-1}\boldsymbol{X}_2^\top \boldsymbol{W}\boldsymbol{X}_1]\boldsymbol{\epsilon}$ and, after some algebra, the coefficients b_ks can be expressed in matrix notation as

$$\begin{aligned}
b_1 &= \frac{1}{2}\text{tr}\{\boldsymbol{W}'\boldsymbol{E}\boldsymbol{B}^{(2)} - (\boldsymbol{W} + 2\boldsymbol{W}')\boldsymbol{B}^{(3)}\} \\
&\quad - \frac{1}{4}\text{tr}\{\boldsymbol{W}\boldsymbol{Z}_d\boldsymbol{B} + (\boldsymbol{W} + 4\boldsymbol{W}')\boldsymbol{Z}_{2d}\boldsymbol{B}\}, \\
b_2 &= \frac{1}{4}\text{tr}\{\boldsymbol{W}(\boldsymbol{Z}_d - \boldsymbol{Z}_{2d})\boldsymbol{B} + (\boldsymbol{W} + 2\boldsymbol{W}')\boldsymbol{B}^{(3)}\}, \\
b_3 &= \frac{1}{12}\text{tr}\{\boldsymbol{W}\boldsymbol{B}^{(3)}\},
\end{aligned}$$

and $b_0 = -(b_1 + b_2 + b_3)$. The b_ks are of order $O(n^{-1/2})$ and all quantities, except $\boldsymbol{\epsilon}$, are evaluated under the null hypothesis $\mathcal{H}_0 : \boldsymbol{\beta}_1 = \boldsymbol{\beta}_{10}$.

Notice that the b_ks depend on the particular model matrix in question through the matrix $\boldsymbol{Z}_d$. They involve the matrices $\boldsymbol{W}$ and $\boldsymbol{W}'$, which determine the censoring mechanism. For example, for generalized type I censoring we have $\boldsymbol{W} = \text{diag}\{1 - \mathrm{e}^{-\rho_1}, \ldots, 1 - \mathrm{e}^{-\rho_n}\}$ and $\boldsymbol{W}' = \text{diag}\{-\rho_1\mathrm{e}^{-\rho_1}, \ldots, -\rho_n\mathrm{e}^{-\rho_n}\}$ with $\rho_l = c_l/\theta_l = c_l\mathrm{e}^{-\mu_l}$, whereas for type II censoring these matrices become $\boldsymbol{W} = (r/n)\boldsymbol{I}_n$ and $\boldsymbol{W}' = \boldsymbol{0}_{n,n}$. The b_ks are also functions of the unknown means. Unfortunately, these coefficients are very difficult to interpret.

The coefficients b_ks can be simplified by examining special cases. For example, no censoring means that c_l ($l = 1, \ldots, n$) goes to infinity for generalized type I censoring, and $r = n$ for type II censoring. For type II censoring we have $\boldsymbol{W} = \boldsymbol{I}_n$ and $\boldsymbol{W}' = \boldsymbol{0}_{n,n}$, and hence the coefficients reduce to

$$b_1 = -\frac{1}{2}\mathrm{tr}\{\boldsymbol{B}^{(3)}\} - \frac{1}{4}\mathrm{tr}\{(\boldsymbol{Z}_d + \boldsymbol{Z}_{2d})\boldsymbol{B}\},$$
$$b_2 = \frac{1}{4}\mathrm{tr}\{(\boldsymbol{Z}_d - \boldsymbol{Z}_{2d})\boldsymbol{B} + \boldsymbol{B}^{(3)}\}, \quad b_3 = \frac{1}{12}\mathrm{tr}\{\boldsymbol{B}^{(3)}\},$$

and $b_0 = -(b_1 + b_2 + b_3)$. For testing the null hypothesis $\mathcal{H}_0 : \boldsymbol{\beta} = \boldsymbol{\beta}_0$ (ie, $q = p$), the coefficients above take the form

$$b_0 = \frac{1}{6}\mathrm{tr}\{\boldsymbol{B}^{(3)}\}, \quad b_1 = -\frac{1}{4}\mathrm{tr}\{2\boldsymbol{B}^{(3)} + \boldsymbol{Z}_d\boldsymbol{B}\},$$
$$b_2 = \frac{1}{4}\mathrm{tr}\{\boldsymbol{Z}_d\boldsymbol{B} + \boldsymbol{B}^{(3)}\}, \quad b_3 = \frac{1}{12}\mathrm{tr}\{\boldsymbol{B}^{(3)}\}.$$

Also, for the simple regression model $y_l = \beta x_l + \varepsilon_l$, where ε_l has a standard extreme value distribution and the null hypothesis is $\mathcal{H}_0 : \beta = \beta_0$, the coefficients above reduce to $b_0 = \epsilon^3 h_2/6$, $b_1 = -\epsilon^3 h_2/2 - \epsilon h_2/(4h_1)$, $b_2 = \epsilon h_2/(4h_1) + \epsilon^3 h_2/4$, $b_3 = \epsilon^3 h_2/12$, where $\epsilon = \beta - \beta_0$ is assumed to be of order $O(n^{-1/2})$, $h_1 = \sum_{l=1}^{n} x_l^2$, and $h_2 = \sum_{l=1}^{n} x_l^3$. Further, if $x_l = 1$ for all $l = 1, \ldots, n$, then $h_1 = h_2 = n$ and the coefficients follow. It corresponds to the case where the observations are independent and identically distributed.

For type II censoring and for testing $\mathcal{H}_0 : \boldsymbol{\beta}_1 = \boldsymbol{\beta}_{10}$, the coefficients that define the nonnull cumulative distribution function of the gradient statistic are

$$b_1 = -\frac{r}{2n}\mathrm{tr}\{\boldsymbol{B}^{(3)}\} - \frac{r}{4n}\mathrm{tr}\{(\boldsymbol{Z}_d + \boldsymbol{Z}_{2d})\boldsymbol{B}\},$$
$$b_2 = \frac{r}{4n}\mathrm{tr}\{(\boldsymbol{Z}_d - \boldsymbol{Z}_{2d})\boldsymbol{B} + \boldsymbol{B}^{(3)}\}, \quad b_3 = \frac{r}{12n}\mathrm{tr}\{\boldsymbol{B}^{(3)}\},$$

and $b_0 = -(b_1 + b_2 + b_3)$. Finally, for $\mathcal{H}_0 : \beta = \beta_0$ and under the simple regression model $y_l = \beta x_l + \varepsilon_l$, the coefficients above take the form $b_0 = r\epsilon^3 h_2/(6n)$, $b_1 = -r\epsilon^3 h_2/(2n) - r\epsilon h_2/(4nh_1)$, $b_2 = r\epsilon h_2/(4nh_1) + r\epsilon^3 h_2/(4n)$, and $b_3 = r\epsilon^3 h_2/(12n)$, where ϵ, h_1, and h_2 were defined before.

Now, denote by Π_i the power function, up to order $O(n^{-1/2})$, of the test that uses the statistic S_i, for $i = \mathrm{LR}, \mathrm{T}$. We have that

$$\Pi_{\mathrm{LR}} - \Pi_{\mathrm{T}} = k_1 g_{q+4,\lambda}(x) + k_2 g_{q+6,\lambda}(x),$$

where x is replaced by the appropriate quantile of the reference distribution for the chosen nominal level, and

$$k_1 = -\frac{1}{2}\mathrm{tr}\{\boldsymbol{W}(\boldsymbol{Z}_d - \boldsymbol{Z}_{2d})\boldsymbol{B}\}, \quad k_2 = -\frac{1}{6}\mathrm{tr}\{\boldsymbol{W}\boldsymbol{B}^{(3)}\}.$$

We arrive at the following general conclusion: $\Pi_{\text{LR}} > \Pi_{\text{T}}$ if $k_1 \geq 0$ and $k_2 \geq 0$ with $k_1 + k_2 > 0$, and $\Pi_{\text{LR}} < \Pi_{\text{T}}$ if $k_1 \leq 0$ and $k_2 \leq 0$ with $k_1 + k_2 < 0$.

Finally, consider the simple regression model

$$y_l = \beta x_l + \varepsilon_l, \quad l = 1, \ldots, n,$$

under no censoring. The null hypothesis of interest is $\mathcal{H}_0 : \beta = \beta_0$. Under this null hypothesis, the local power functions up to order $O(n^{-1/2})$ of the LR and gradient tests can be expressed as

$$\Pi_i = 1 - \left[G_{1,\lambda}(x) + \sum_{k=0}^{3} b_{ik} G_{1+2k,\lambda}(x) \right], \quad i = \text{LR}, \text{T}, \tag{2.14}$$

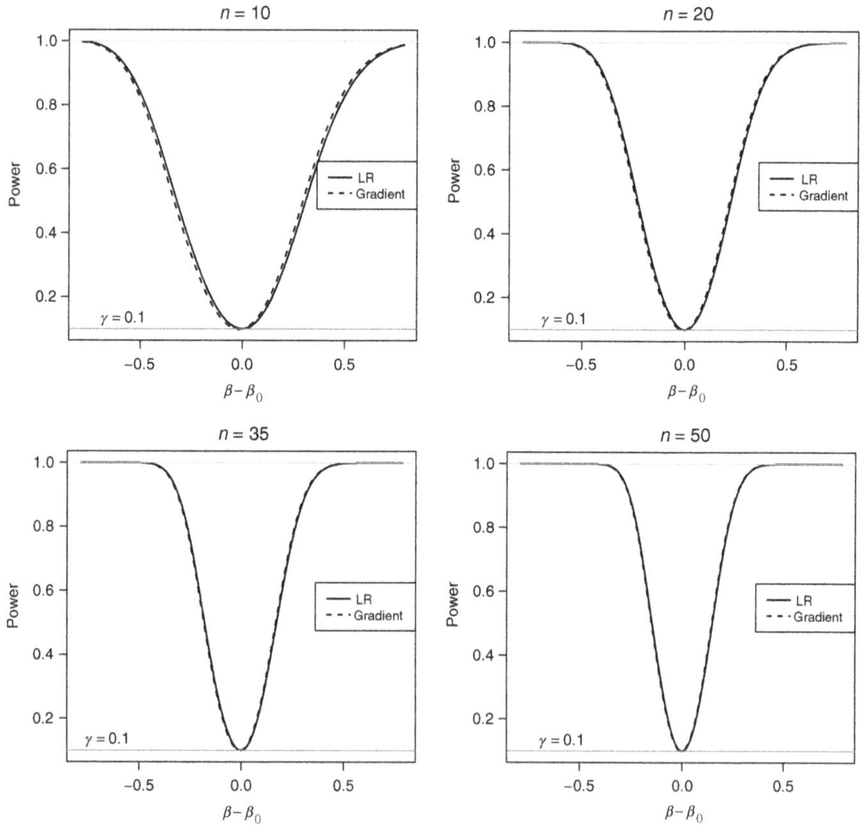

Fig. 2.3 Power functions up to order $O(n^{-1/2})$ for some values of $\epsilon = \beta - \beta_0$.

where $\lambda = (1/2)\epsilon^2 h_1$, $b_{T0} = \epsilon^3 h_2/6$, $b_{T1} = -\epsilon^3 h_2/2 - \epsilon h_2/(4h_1)$, $b_{T2} = \epsilon h_2/(4h_1) + \epsilon^3 h_2/4$, $b_{T3} = \epsilon^3 h_2/12$, $b_{LR2} = 2b_{LR0} = -(2/3)b_{LR1} = \epsilon^3 h_2/3$, $b_{LR3} = 0$. Also, ϵ, h_1, and h_2 were defined before. We shall compare the local powers of the LR and gradient tests for testing $\mathcal{H}_0$ by considering different values for $\epsilon = \beta - \beta_0$ (close to the null hypothesis) and for some sample sizes. We set $\beta_0 = 0$, $n = 10, 20, 35$, and 50, and $\gamma = 10\%$. The values of the covariate were selected as random draws from the uniform $\mathcal{U}(1, 2)$ distribution without loss of generality. The local powers of the LR and gradient tests obtained from the power functions up to order $O(n^{-1/2})$ are displayed in Fig. 2.3. As can be seen from this figure, in general, none of the two tests is uniformly most powerful for testing the null hypothesis $\mathcal{H}_0 : \beta = 0$. Additionally, as the sample size grows, the powers of the two tests increase and the power curves of the tests become indistinguishable from each other, as expected.

CHAPTER 3

The Bartlett-Corrected Gradient Statistic

3.1 INTRODUCTION

In Chapter 2, it was showed that the gradient test can be an interesting alternative to the classic large-sample tests, namely the LR, Wald, and Rao score tests, since none is uniformly superior to the others in terms of second-order local power. Additionally, as remarked before, the gradient test statistic does not require one to obtain, estimate, or invert an information matrix, unlike the Wald and score statistics. Like the classic large-sample tests, the gradient test is performed using approximate critical values. Such critical values are obtained from the test statistic limiting distribution when the null hypothesis is true. The approximation holds when the number of observations in the sample tends to infinity, and it is thus expected to deliver reliable inferences in large samples.

The exact null distribution (ie, the exact distribution under the null hypothesis) of the gradient statistic is usually unknown and the test relies

The Gradient Test. http://dx.doi.org/10.1016/B978-0-12-803596-2.00003-X

upon an asymptotic approximation. The χ^2 distribution is used as a large-sample approximation to the true null distribution (ie, the true distribution under the null hypothesis) of this statistic. However, for small sample sizes, the χ^2 distribution may be a poor approximation to the true null distribution and the asymptotic approximation may deliver inaccurate inference; that is, when the sample is not large, size distortions are likely to arise and the effective type I error probability may not be close to the nominal size selected by the practitioner. In order to overcome this shortcoming, an alternative strategy is to use a higher-order asymptotic theory.

The main result in Lemonte and Ferrari [12] regarding the local power of the gradient test up to an error of order $O(n^{-1})$ represents the first step in the study of higher-order asymptotic properties of the gradient test. The usual route for deriving expansions for the distribution function of asymptotic χ^2 test statistics involves multivariate Edgeworth series expansions. Although such a route has been followed by many authors, it is extremely lengthy and tedious. Recently, Vargas et al. [19] demonstrated how to improve the χ^2 approximation for the exact distribution of the gradient statistic in wide generality by multiplying it by a Bartlett-type correction factor by using a Bayesian route based on a shrinkage argument originally suggested by Ghosh and Mukerjee [20]. Although it uses a Bayesian approach, this technique can be used to solve frequentist problems, such as the derivation of Bartlett corrections and tail probabilities. Under the null hypothesis, the Bartlett-corrected gradient statistic is distributed as χ^2 up to an error of order $O(n^{-3/2})$, whereas the uncorrected gradient statistic has a χ^2 distribution up to an error of order $O(n^{-1})$; that is, the Bartlett-type correction makes the approximation error be reduced from $O(n^{-1})$ to $O(n^{-3/2})$ and therefore the Bartlett-corrected gradient test is expected to display superior finite sample behavior than the usual gradient test.

3.2 NULL DISTRIBUTION UP TO ORDER $O(n^{-1})$

Let $\ell(\boldsymbol{\theta})$ be the log-likelihood function. Suppose the interest lies in testing the composite null hypothesis $\mathcal{H}_0 : \boldsymbol{\theta}_1 = \boldsymbol{\theta}_{10}$ against $\mathcal{H}_a : \boldsymbol{\theta}_1 \neq \boldsymbol{\theta}_{10}$, where $\boldsymbol{\theta} = (\boldsymbol{\theta}_1^\top, \boldsymbol{\theta}_2^\top)^\top$, $\boldsymbol{\theta}_1$ and $\boldsymbol{\theta}_2$ are parameter vectors of dimensions q and $p - q$, respectively, and $\boldsymbol{\theta}_{10}$ is a q-dimensional fixed vector. The gradient statistic for testing the null hypothesis $\mathcal{H}_0 : \boldsymbol{\theta}_1 = \boldsymbol{\theta}_{10}$ is defined in Eq. (1.3).

Let us introduce some notation. Let $D_j = \partial/\partial\theta_j$ $(j = 1, \ldots, p)$ be the differential operator. We also define $U_j = D_j\ell(\boldsymbol{\theta})$, $U_{jr} = D_jD_r\ell(\boldsymbol{\theta})$, $U_{jrs} = D_jD_rD_s\ell(\boldsymbol{\theta})$, and $U_{jrsu} = D_jD_rD_sD_u\ell(\boldsymbol{\theta})$. We make the same assumptions, such as the regularity of the first four derivatives of $\ell(\boldsymbol{\theta})$ with respect to $\boldsymbol{\theta}$ and the existence and uniqueness of the MLE of $\boldsymbol{\theta}$, as those fully outlined by Hayakawa [21]. Let $\kappa_{jr} = \mathbb{E}(U_{jr})$, $\kappa_{jrs} = \mathbb{E}(U_{jrs})$, $\kappa_{jrsu} = \mathbb{E}(U_{jrsu})$, $\kappa_{jr}^{(s)} = D_s\kappa_{jr}$, $\kappa_{jr}^{(su)} = D_sD_u\kappa_{jr}$, and $\kappa_{jrs}^{(u)} = D_u\kappa_{jrs}$. Further, let $\boldsymbol{K}$ be the per observation Fisher information matrix

$$\boldsymbol{K} = -((\kappa_{jr}))_{j,r=1,\dots,p} = \begin{bmatrix} \boldsymbol{K}_{11} & \boldsymbol{K}_{12} \\ \boldsymbol{K}_{21} & \boldsymbol{K}_{22} \end{bmatrix},$$

with $\boldsymbol{K}^{-1} = -((\kappa^{jr}))$ denoting its inverse. Finally, define the matrices

$$\boldsymbol{A} = ((a^{jr}))_{j,r=1,\dots,p} = \begin{bmatrix} \boldsymbol{0}_{q,q} & \boldsymbol{0}_{q,p-q} \\ \boldsymbol{0}_{p-q,q} & \boldsymbol{K}_{22}^{-1} \end{bmatrix},$$

$$\boldsymbol{M} = ((m^{jr}))_{j,r=1,\dots,p} = \boldsymbol{K}^{-1} - \boldsymbol{A}.$$

In what follows, we use the Einstein summation convention, where $\sum'$ denotes summation over all components of $\boldsymbol{\theta}$; that is, the indices j, r, s, k, l, and u range over 1 to p. We have the following theorem.

Theorem 3.1. *The asymptotic expansion for the null cumulative distribution function of the gradient statistic for testing $\mathcal{H}_0 : \boldsymbol{\theta}_1 = \boldsymbol{\theta}_{10}$ against $\mathcal{H}_a : \boldsymbol{\theta}_1 \neq \boldsymbol{\theta}_{10}$ is*

$$\Pr(S_T \le x|\mathcal{H}_0) = G_q(x) + \frac{1}{24n}\sum_{i=0}^{3} R_i G_{q+2i}(x) + O(n^{-3/2}), \qquad (3.1)$$

where $G_z(\cdot)$ is the cumulative distribution function of a (central) χ^2 random variable with z degrees of freedom, $R_1 = 3A_3 - 2A_2 + A_1$, $R_2 = A_2 - 3A_3$, $R_3 = A_3$, $R_0 = -(R_1 + R_2 + R_3)$, and

$$\begin{aligned} A_1 = 3{\sum}' \kappa_{jrs}\kappa_{klu}\Big[& m^{jr}a^{lu}(m^{sk} + 2a^{sk}) \\ & + a^{jr}m^{sk}a^{lu} + 2m^{jk}a^{rl}a^{su}\Big] \\ - 12{\sum}' \kappa_{jr}^{(s)}\kappa_{kl}^{(u)}\Big(& \kappa^{sj}\kappa^{rk}\kappa^{lu} + a^{sj}a^{rk}a^{lu} \\ & + \kappa^{sk}\kappa^{lj}\kappa^{ru} + a^{sk}a^{lj}a^{ru}\Big) \\ - 6{\sum}' \kappa_{jrs}\kappa_{kl}^{(u)}\Big[& \left(a^{su} - \kappa^{su}\right)\left(\kappa^{jk}\kappa^{lr} - a^{jk}a^{lr}\right) \\ & + 2a^{rs}\left(\kappa^{jk}\kappa^{lu} - a^{jk}a^{lu}\right) + 2a^{rk}a^{ls}m^{ju} \\ & + m^{jr}\left(a^{sk}a^{lu} + \kappa^{sk}\kappa^{lu}\right)\Big] \end{aligned}$$

$$- 6 \sum{}' \kappa_{jrs}^{(u)} \left[m^{jr} \left(a^{su} - \kappa^{su} \right) + 2 m^{ju} a^{rs} \right]$$
$$+ 6 \sum{}' \kappa_{jrsu} m^{jr} a^{su} + 12 \sum{}' \kappa_{rs}^{(ju)} \left(\kappa^{jr} \kappa^{su} - a^{jr} a^{su} \right),$$

$$A_2 = -3 \sum{}' \kappa_{jrs} \kappa_{klu} \Big[m^{jr} m^{sk} a^{lu} + m^{jr} a^{sk} m^{lu} + 2 m^{jk} m^{rl} a^{su}$$
$$+ \frac{1}{4} \left(3 m^{jr} m^{sk} m^{lu} + 2 m^{jk} m^{rl} m^{su} \right) \Big]$$
$$+ 6 \sum{}' \kappa_{jrs} \kappa_{kl}^{(u)} \Big[m^{su} \left(\kappa^{jk} \kappa^{lr} - a^{jk} a^{lr} \right)$$
$$+ m^{jr} \left(\kappa^{sk} \kappa^{lu} - a^{sk} a^{lu} \right) \Big]$$
$$+ 6 \sum{}' \kappa_{jrs}^{(u)} m^{jr} m^{su} - 3 \sum{}' \kappa_{jrsu} m^{jr} m^{su},$$
$$A_3 = \frac{1}{4} \sum{}' \kappa_{jrs} \kappa_{klu} \left(3 m^{jr} m^{sk} m^{lu} + 2 m^{jk} m^{rl} m^{su} \right).$$

Proof. Readers are referred to Vargas et al. [19]. □

If the null hypothesis is simple, then $q = p$, $\boldsymbol{A} = \boldsymbol{0}_{p,p}$ and $\boldsymbol{M} = \boldsymbol{K}^{-1}$. Therefore, an immediate consequence of Theorem 3.1 is the following corollary.

Corollary 3.1. *The asymptotic expansion for the null cumulative distribution function of the gradient statistic for testing* $\mathcal{H}_0 : \boldsymbol{\theta} = \boldsymbol{\theta}_0$ *against* $\mathcal{H}_a : \boldsymbol{\theta} \neq \boldsymbol{\theta}_0$ *is given by Eq. (3.1) with* $q = p$, $R_1 = 3A_3 - 2A_2 + A_1$, $R_2 = A_2 - 3A_3$, $R_3 = A_3$, $R_0 = -(R_1 + R_2 + R_3)$, *and*

$$A_1 = -12 \sum{}' \kappa_{jr}^{(s)} \kappa_{kl}^{(u)} \left(\kappa^{sj} \kappa^{rk} \kappa^{lu} + \kappa^{sk} \kappa^{lj} \kappa^{ru} \right)$$
$$+ 6 \sum{}' \kappa_{jrs} \kappa_{kl}^{(u)} \left(\kappa^{su} \kappa^{jk} \kappa^{lr} + \kappa^{jr} \kappa^{sk} \kappa^{lu} \right)$$
$$+ 12 \sum{}' \kappa_{rs}^{(ju)} \kappa^{jr} \kappa^{su} - 6 \sum{}' \kappa_{jrs}^{(u)} \kappa^{jr} \kappa^{su},$$

$$A_2 = \frac{3}{4}\sum{}' \kappa_{jrs}\kappa_{klu}(3\kappa^{jr}\kappa^{sk}\kappa^{lu} + 2\kappa^{jk}\kappa^{rl}\kappa^{su})$$
$$- 6\sum{}' \kappa_{jrs}\kappa_{kl}^{(u)}\left(\kappa^{su}\kappa^{jk}\kappa^{lr} + \kappa^{jr}\kappa^{sk}\kappa^{lu}\right)$$
$$+ 6\sum{}' \kappa_{jrs}^{(u)}\kappa^{jr}\kappa^{su} - 3\sum{}' \kappa_{jrsu}\kappa^{jr}\kappa^{su},$$
$$A_3 = -\frac{1}{4}\sum{}' \kappa_{jrs}\kappa_{klu}\left(3\kappa^{jr}\kappa^{sk}\kappa^{lu} + 2\kappa^{jk}\kappa^{rl}\kappa^{su}\right).$$

Next, we present a Bartlett-type corrected gradient statistic. A Bartlett-type correction is a multiplying factor, which depends on the statistic itself, that results in a modified statistic that follows a χ^2 distribution with approximation error of order less than $O(n^{-1})$. Cordeiro and Ferrari [22] obtained a general formula for a Bartlett-type correction for a wide class of statistics that have a χ^2 distribution asymptotically. A special case is when the cumulative distribution function of the statistic can be written as Eq. (3.1), independently of the coefficients R_1, R_2, and R_3. Hence, from Theorem 3.1 and the results in Cordeiro and Ferrari [22], we have the following corollary.

Corollary 3.2. *The modified gradient statistic*

$$S_{\mathrm{T}}^{*} = S_{\mathrm{T}}\left[1 - \left(c + bS_{\mathrm{T}} + aS_{\mathrm{T}}^{2}\right)\right], \tag{3.2}$$

where

$$a = \frac{A_3}{12nq(q+2)(q+4)}, \quad b = \frac{A_2 - 2A_3}{12nq(q+2)}, \quad c = \frac{A_1 - A_2 + A_3}{12nq},$$

has a χ_q^2 distribution up to an error of order $O(n^{-3/2})$ under the null hypothesis.

The factor $[1-(c+bS_{\mathrm{T}}+aS_{\mathrm{T}}^{2})]$ in Eq. (3.2) can be regarded as a Bartlett-type correction factor for the gradient statistic in such a way that the null distribution of S_{T}^{*} is better approximated by the reference χ^2 distribution than the distribution of the uncorrected gradient statistic S_{T}.

Instead of modifying the test statistic as in Eq. (3.2), we may modify the reference χ^2 distribution using the inverse expansion formula in Hill and Davis [23]. To be specific, let $\gamma \in (0, 1)$ be the desired level of the test, and $x_{1-\gamma}$ be the $1-\gamma$ percentile of the χ^2 limiting distribution of the test statistic. From expansion (3.1), we have the following corollary.

Corollary 3.3. *The asymptotic expansion for the* $1-\gamma$ *percentile of the gradient statistic* S_{T} *takes the form*

$$\begin{aligned} z_{1-\gamma} = x_{1-\gamma} &+ \frac{1}{12n}\left\{\frac{A_3 x_{1-\gamma}}{q(q+2)(q+4)}\left[x_{1-\gamma}^2 + (q+4)x_{1-\gamma}\right.\right. \\ &\left. +(q+2)(q+4)\right] + \frac{x_{1-\gamma}(x_{1-\gamma}+q+2)}{q(q+2)}(A_2 - 3A_3) \\ &\left. +\frac{x_{1-\gamma}}{q}(3A_3 - 2A_2 + A_1)\right\} + O(n^{-3/2}), \end{aligned} \tag{3.3}$$

where $\Pr(\chi_q^2 \geq x_{1-\gamma}) = \gamma$.

In general, Eqs. (3.2) and (3.3) depend on unknown parameters. In this case, we can replace these unknown parameters by their MLEs obtained under the null hypothesis. The approximation order remains the same when one does so. It should be noticed that the improved gradient test of the null hypothesis $\mathcal{H}_0$ may be performed in three ways: (i) by referring the modified gradient statistic S_{T}^* in Eq. (3.2) to the χ_q^2 distribution; (ii) by referring the gradient statistic S_{T} to the approximate cumulative distribution function (3.1); or (iii) by comparing S_{T} with the modified upper percentile in Eq. (3.3). These three procedures are equivalent to order $O(n^{-1})$.

Finally, the three moments of the gradient statistic, up to order $O(n^{-1})$ and under the null hypothesis, are presented in the following corollary.

Corollary 3.4. *The mean, variance, and third central moment of the gradient statistic, up to order* $O(n^{-1})$ *under the null hypothesis, are*

$$\mathbb{E}(S_{\mathrm{T}}) = q + \frac{A_1}{12n}, \quad \mathbb{VAR}(S_{\mathrm{T}}) = 2q + \frac{A_1 + A_2}{3n},$$

$$\mu_3(S_{\mathrm{T}}) = 8q + \frac{2(A_1 + 2A_2 + A_3)}{n},$$

respectively.

3.3 THE ONE-PARAMETER CASE

We initially assume that the model is indexed by a scalar unknown parameter, say ϕ. The interest lies in testing the null hypothesis $\mathcal{H}_0 : \phi = \phi_0$ against $\mathcal{H}_a : \phi \neq \phi_0$, where ϕ_0 is a fixed value. Let $\kappa_{\phi\phi} = \mathbb{E}(\partial^2 \ell(\phi)/\partial\phi^2)$, $\kappa_{\phi\phi\phi} = \mathbb{E}(\partial^3 \ell(\phi)/\partial\phi^3)$, $\kappa_{\phi\phi\phi\phi} = \mathbb{E}(\partial^4 \ell(\phi)/\partial\phi^4)$, $\kappa_{\phi\phi}^{(\phi)} = \partial\kappa_{\phi\phi}/\partial\phi$, $\kappa_{\phi\phi\phi}^{(\phi)} = \partial\kappa_{\phi\phi\phi}/\partial\phi$, and $\kappa_{\phi\phi}^{(\phi\phi)} = \partial^2\kappa_{\phi\phi}/\partial\phi^2$. The gradient statistic for testing $\mathcal{H}_0$ is $S_{\mathrm{T}} = U(\phi_0)(\widehat{\phi} - \phi_0)$, where $U(\phi)$ is the score function of ϕ, and $\hat{\phi}$ is the MLE of ϕ. Here, A_1, A_2, and A_3 given in Corollary 3.1 reduce to

$$A_1 = \frac{6\kappa_{\phi\phi}(2\kappa_{\phi\phi}^{(\phi\phi)} - \kappa_{\phi\phi\phi}^{(\phi)}) + 12\kappa_{\phi\phi}^{(\phi)}(\kappa_{\phi\phi\phi} - 2\kappa_{\phi\phi}^{(\phi)})}{\kappa_{\phi\phi}^3}, \tag{3.4}$$

$$A_2 = \frac{12\kappa_{\phi\phi}(2\kappa_{\phi\phi\phi}^{(\phi)} - 3\kappa_{\phi\phi\phi\phi}) + 3\kappa_{\phi\phi\phi}(5\kappa_{\phi\phi\phi} - 16\kappa_{\phi\phi}^{(\phi)})}{4\kappa_{\phi\phi}^3}, \tag{3.5}$$

$$A_3 = -\frac{5\kappa_{\phi\phi\phi}^2}{4\kappa_{\phi\phi}^3}. \tag{3.6}$$

We now present some examples.

Example 3.1 (Exponential distribution). Let $x_1, \ldots, x_n$ be a sample of size n from an exponential distribution with density

$$f(x;\phi) = \frac{1}{\phi}\exp(-x/\phi), \quad x > 0, \quad \phi > 0.$$

Here, $\kappa_{\phi\phi} = -\phi^{-2}$, $\kappa_{\phi\phi\phi} = 4\phi^{-3}$, and $\kappa_{\phi\phi\phi\phi} = -18\phi^{-4}$. The gradient statistic assumes the form $S_{\mathrm{T}} = n(\bar{x} - \phi_0)^2/\phi_0^2$, where $\bar{x} = n^{-1}\sum_{l=1}^{n} x_l$. It is easy to see that $A_1 = 0$, $A_2 = 18$, and $A_3 = 20$. The first three moments (up to order $O(n^{-1})$) of S_{T} are $\mathbb{E}(S_{\mathrm{T}}) = 1$, $\mathbb{VAR}(S_{\mathrm{T}}) = 2 + 6/n$, and $\mu_3(S_{\mathrm{T}}) = 8 + 112/n$. A partial verification of these results can be accomplished by comparing the exact moments of S_{T} with the approximate moments given above. Since $n\bar{x}$ has a gamma distribution with parameters n and $1/(n\phi)$, it can be shown that the first three exact moments of S_{T} are 1, $2 + 6/n$, and $8 + 112/n + 120/n^2$, respectively. These moments differ from the approximate moments obtained from Corollary 3.4 only in terms of order less than $O(n^{-1})$. The Bartlett-type corrected gradient statistic obtained from Corollary 3.2 is

$$S_T^* = S_T\left[1 - \frac{3 - 11S_T + 2S_T^2}{18n}\right].$$

Example 3.2 (One-parameter exponential family). Let $x_1, \ldots, x_n$ be n independent observations in which each x_l has a distribution in the one-parameter exponential family with probability density function

$$f(x; \phi) = \frac{1}{\xi(\phi)} \exp[-\alpha(\phi)d(x) + v(x)],$$

where $\alpha(\cdot)$, $v(\cdot)$, $d(\cdot)$ and $\xi(\cdot)$ are known functions, and $\beta(\phi) = \xi'(\phi)/(\xi(\phi)\alpha'(\phi))$. Here, primes denote derivatives with respect to ϕ. We have that $\kappa_{\phi\phi} = -\alpha'\beta'$, $\kappa_{\phi\phi\phi} = -(2\alpha''\beta' + \alpha'\beta'')$, and $\kappa_{\phi\phi\phi\phi} = -3\alpha''\beta'' - 3\alpha'''\beta' - \alpha'\beta'''$. The gradient statistic takes the form $S_T = n(\phi_0 - \hat{\phi})\alpha'(\phi_0)[\beta(\phi_0) + \bar{d}]$, where $\bar{d} = n^{-1}\sum_{l=1}^{n} d(x_l)$. Using Eqs. (3.4)–(3.6), we can write

$$A_1 = \frac{6}{\alpha'\beta'}\left[2\left(\frac{\beta''}{\beta'}\right)^2 + \frac{\alpha''\beta''}{\alpha'\beta'} - \frac{\beta'''}{\beta'}\right], \quad A_3 = \frac{5}{\alpha'\beta'}\left(\frac{\alpha''}{\alpha'} + \frac{\beta''}{2\beta'}\right)^2,$$

$$A_2 = \frac{3}{\alpha'\beta'}\left\{\frac{\beta''}{\beta'}\left(\frac{4\alpha''}{\alpha'} - \frac{\beta''}{4\beta'}\right) + 3\left[\left(\frac{\alpha''}{\alpha'}\right)^2 + \left(\frac{\beta''}{\beta'}\right)^2\right]\right\} - \frac{3}{\alpha'\beta'}\left(\frac{\alpha'''}{\alpha'} - \frac{\beta'''}{\beta'}\right).$$

We now present some special cases:

1. Normal ($\phi > 0$, $\mu \in \mathbb{R}$, $x \in \mathbb{R}$):
   - μ known: We have $A_1 = 0$, $A_2 = 36$, and $A_3 = 40$. The first three moments of S_T up to order $O(n^{-1})$ are $\mathbb{E}(S_T) = 1$, $\mathbb{VAR}(S_T) = 2(1+6/n)$, and $\mu_3(S_T) = 8(1+29/n)$. The Bartlett-corrected gradient statistic is

     $$S_T^* = S_T\left[1 - \frac{1 - 11S_T/3 + 2S_T^2/3}{3n}\right].$$

   - ϕ known: Here, $A_1 = A_2 = A_3 = 0$, as expected.
2. Inverse normal ($\phi > 0$, $\mu > 0$, $x > 0$):
   - μ known: Here, $A_1 = 24$, $A_2 = 30$, and $A_3 = 10$, and the three first moments of S_T are $\mathbb{E}(S_T) = 1 + 2/n$, $\mathbb{VAR}(S_T) = 2 + 18/n$, and

$\mu_3(S_T) = 8 + 188/n$. The Bartlett-corrected gradient statistic takes the form

$$S_T^* = S_T\left[1 - \frac{(S_T+2)(S_T+3)}{18n}\right].$$

- ϕ known: We have $A_1 = 0$ and $A_2 = A_3 = 45\mu/\phi$. The first three approximate moments of S_T are $\mathbb{E}(S_T) = 1$, $\mathbb{VAR}(S_T) = 2 + 15\mu/(n\phi)$, and $\mu_3(S_T) = 8 + 270\mu/(n\phi)$. Also,

$$S_T^* = S_T\left[1 - \frac{\mu S_T(S_T-5)}{4n\phi}\right].$$

3. Truncated extreme value ($\phi > 0, x > 0$). We have $A_1 = 0, A_2 = 12$, $A_3 = 20$, $\mathbb{E}(S_T) = 1$, $\mathbb{VAR}(S_T) = 2 + 4/n$, $\mu_3(S_T) = 8 + 88/n$, and

$$S_T^* = S_T\left[1 - \frac{12 - 15S_T + 2S_T^2}{18n}\right].$$

4. Pareto ($\phi > 0, k > 0, k$ known, $x > k$). Here, $A_1 = 12, A_2 = 15, A_3 = 5$, $\mathbb{E}(S_T) = 1 + 1/n$, $\mathbb{VAR}(S_T) = 2 + 9/n$, $\mu_3(S_T) = 8 + 94/n$, and

$$S_T^* = S_T\left[1 - \frac{(S_T+2)(S_T+3)}{36n}\right].$$

5. Power ($\theta > 0, \phi > 0, \theta$ known, $x > \theta$). The As, the first three approximate moments, and the Bartlett-type corrected statistic coincide with those obtained for the Pareto distribution.
6. Laplace ($\theta > 0, k \in \mathbb{R}, k$ known, $x \in \mathbb{R}$). We have $A_1 = 0, A_2 = 18$, $A_3 = 20$, $\mathbb{E}(S_T) = 1$, $\mathbb{VAR}(S_T) = 2 + 6/n$, $\mu_3(S_T) = 8 + 112/n$, and

$$S_T^* = S_T\left[1 - \frac{3 - 11S_T + 2S_T^2}{18n}\right].$$

3.4 MODELS WITH TWO ORTHOGONAL PARAMETERS

The two-parameter families of distributions under orthogonality of the parameters, say ϕ and η, will be the subject of this section; that is, we impose

$$\mathbb{E}\left(\frac{\partial^2 \ell(\phi,\eta)}{\partial\phi\partial\eta}\right) = 0.$$

The null hypothesis under test is $\mathcal{H}_0 : \phi = \phi_0$, where ϕ_0 is a fixed value, and η acts as a nuisance parameter. The orthogonality between ϕ and η leads to considerable simplification in the formulas of A_1, A_2, and A_3. Here,

$\kappa_{\phi\phi\eta} = \mathbb{E}(\partial^3\ell(\phi,\eta)/\partial\eta\partial\phi^2)$, $\kappa_{\phi\phi\eta}^{(\eta)} = \partial\kappa_{\phi\phi\eta}/\partial\eta$, etc. After some algebra, we have

$$A_1 = A_{1\phi} + A_{1\phi\eta}, \quad A_2 = A_{2\phi} + A_{2\phi\eta}, \quad A_3 = -\frac{5\kappa_{\phi\phi\phi}^2}{4\kappa_{\phi\phi}^3}, \tag{3.7}$$

where $A_{1\phi}$ and $A_{2\phi}$ are equal to A_1 and A_2 given in Eqs. (3.4) and (3.5), respectively, and

$$\begin{aligned} A_{1\phi\eta} = {} & \frac{3\left[2\kappa_{\phi\phi\eta}\left(2\kappa_{\eta\eta}^{(\eta)} - \kappa_{\eta\eta\eta}\right) + \kappa_{\phi\eta\eta}\left(2\kappa_{\eta\eta}^{(\phi)} - 3\kappa_{\phi\eta\eta}\right)\right]}{\kappa_{\phi\phi}\kappa_{\eta\eta}^2} \\ & + \frac{3\left[4\kappa_{\phi\phi\eta}\kappa_{\phi\phi}^{(\eta)} + \kappa_{\phi\eta\eta}\left(4\kappa_{\phi\phi}^{(\phi)} - \kappa_{\phi\phi\phi}\right)\right]}{\kappa_{\phi\phi}^2\kappa_{\eta\eta}} \\ & + \frac{6\left(\kappa_{\phi\phi\eta\eta} - 2\kappa_{\phi\phi\eta}^{(\eta)} - 2\kappa_{\phi\eta\eta}^{(\phi)}\right)}{\kappa_{\phi\phi}\kappa_{\eta\eta}}, \\ A_{2\phi\eta} = {} & \frac{3\left(3\kappa_{\phi\phi\phi}\kappa_{\phi\eta\eta} + \kappa_{\phi\phi\eta}^2\right)}{\kappa_{\phi\phi}^2\kappa_{\eta\eta}}. \end{aligned}$$

The expressions for $A_{1\phi\eta}$ and $A_{2\phi\eta}$ in Eq. (3.7) can be regarded as the additional contribution introduced in the expansion of the cumulative distribution function of the gradient statistic owing to the fact that η is unknown and has to be estimated from the data.

In the following, we present some examples.

Example 3.3 (Normal distribution). Let $x_1, \ldots, x_n$ be a sample of size n from a normal distribution with mean $\phi \in \mathbb{R}$ and variance $\eta > 0$, both are assumed unknown. The gradient statistic for testing $\mathcal{H}_0 : \phi = \phi_0$ can be written in the form

$$S_T = n\frac{T_1 T_2^{-1}}{1 + T_1 T_2^{-1}},$$

where $T_1 = n(\bar{x}-\phi_0)^2$, $T_2 = \sum_{l=1}^n (x_l-\bar{x})^2$, and $\bar{x} = n^{-1}\sum_{l=1}^n x_l$. Under the null hypothesis, T_1/η and T_2/η are independent with distributions χ_1^2 and χ_{n-1}^2, respectively. It can be shown that $n^{-1}S_T$ follows a beta distribution with parameters $1/2$ and $(n-1)/2$. The first three exact moments of S_T are 1, $2(n-1)/(n+2)$, and $8(n-1)(n-2)/[(n+2)(n+4)]$, respectively. Here, $A_1 = A_3 = 0$ and $A_2 = -18$. The first three approximate moments of S_T are

$\mathbb{E}(S_{\text{T}}) = 1$, $\mathbb{VAR}(S_{\text{T}}) = 2 - 6/n$, and $\mu_3(S_{\text{T}}) = 8 - 72/n$. These moments differ from the exact moments only by terms of order less than $O(n^{-1})$. The Bartlett-type corrected gradient statistic is

$$S_{\text{T}}^{*} = S_{\text{T}}\left[1 - \frac{3 - S_{\text{T}}}{2n}\right].$$

Example 3.4 (Two-parameter Birnbaum-Saunders distribution). The two-parameter Birnbaum-Saunders distribution (see Section 1.5 for more details) has cumulative distribution function in the form $F(x; \phi, \eta) = \Phi(v_x)$, $x > 0$, where $v_x = \phi^{-1}\rho(x/\eta)$, $\rho(z) = z^{1/2} - z^{-1/2}$, and $\Phi(\cdot)$ is the standard normal cumulative distribution function. Here, $\phi > 0$ and $\eta > 0$ are the shape and scale parameters, respectively. We wish to test $\mathcal{H}_0 : \phi = \phi_0$ against the alternative hypothesis $\mathcal{H}_a : \phi \neq \phi_0$, where ϕ_0 is a known positive constant. The gradient statistic to test $\mathcal{H}_0$ is

$$S_{\text{T}} = \frac{n(\hat{\phi} - \phi_0)}{\phi_0^3}[\bar{s} + \bar{r} - (2 + \phi_0^2)],$$

where $\bar{s} = (n\tilde{\eta})^{-1}\sum_{l=1}^{n} x_l$, $\bar{r} = n^{-1}\tilde{\eta}\sum_{l=1}^{n} x_l^{-1}$, and $\tilde{\eta}$ is the restricted MLE of η obtained under $\mathcal{H}_0 : \phi = \phi_0$. We have $\kappa_{\phi\phi} = -2/\phi^2$, $\kappa_{\phi\eta} = 0$, and $\kappa_{\eta\eta} = -[1 + \phi(2\pi)^{-1/2}h(\phi)]/(\phi^2\eta^2)$, where $h(\phi) = \phi(\pi/2)^{1/2} - \pi e^{2/\phi^2}[1 - \Phi(2/\phi)]$. After some algebra, we obtain $A_{1\phi} = -3$, $A_{2\phi} = 69/8$, $A_{2\phi\eta} = -45(2 + \phi^2)/[2\{1 + \phi(2\pi)^{-1/2}h(\phi)\}]$, $A_3 = 125/8$, and

$$A_{1\phi\eta} = \frac{3(2 + \phi^2)}{[1 + \phi(2\pi)^{-1/2}h(\phi)]^2}\left[2(1 + \phi^2) - \frac{(4 + \phi^2)h(\phi)}{\phi\sqrt{2\pi}}\right] + \frac{9 - 15\phi^2/2}{1 + \phi(2\pi)^{-1/2}h(\phi)}.$$

Since the necessary quantities to obtain the As were derived, a Bartlett-corrected gradient statistic may be obtained from Corollary 3.2. It is interesting to note that the As do not depend on the unknown scale parameter η.

In what follows, we shall present a small Monte Carlo simulation study regarding the test of the null hypothesis $\mathcal{H}_0 : \phi = 1$ in Example (3.4). The simulations were performed by setting $\eta = 1$ and sample sizes ranging from 5 to 22 observations. All results are based on 10,000 replications. The size distortions (ie, estimated minus nominal sizes) for the 5% nominal level of the gradient test and its Bartlett-corrected version for different sample sizes are displayed in Fig. 3.1A. It is clear from this figure that the

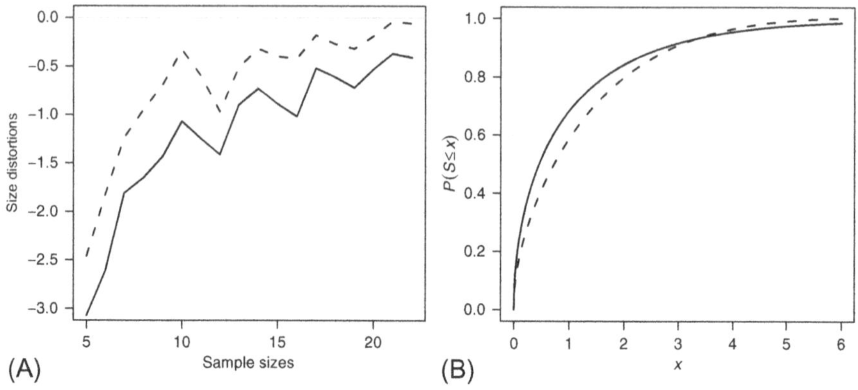

Fig. 3.1 (A) Size distortions of the gradient test (solid) and Bartlett-corrected gradient test (dashes); (B) first-order approximation (solid) and expansion to order $O(n^{-1})$ (dashes) of the null cumulative distribution function of the gradient statistic; Birnbaum-Saunders distribution.

Bartlett-corrected test presents smaller size distortions than the original gradient test. Next, we set $n = 10$ and consider the first-order approximation (χ_1^2 distribution) for the cumulative distribution function of the gradient statistic and the expansion presented in this chapter. Fig. 3.1B displays the curves. The difference between the curves is evident from this figure, and hence, the χ_1^2 distribution may not be a good approximation for the null distribution of the gradient statistic in testing the null hypothesis $\mathcal{H}_0 : \phi = 1$ for the two-parameter Birnbaum-Saunders distribution if the sample is small.

Example 3.5 (Gamma distribution). Let $x_1, \ldots, x_n$ be a sample of size n from a gamma distribution with mean η and coefficient of variation $\phi^{1/2}$. Here, we consider the problem of testing the null hypothesis $\mathcal{H}_0 : (\eta, \phi) = (\eta_0, \phi_0)$, where η_0 and ϕ_0 are fixed positive values. Note that the null hypothesis is simple, and the As can be obtained from Corollary 3.1 with $q = p = 2$. After some algebra, we obtain

$$A_1 = 6(d_1 - d_2 + d_3 - 2d_4),$$

$$A_2 = \frac{18}{\phi} + \frac{3}{4}(d_1 + 2d_2 + 4d_3 - 11d_4),$$

$$A_3 = \frac{20}{\phi} - \frac{1}{4}(9d_1 - 6d_2 + 5d_4),$$

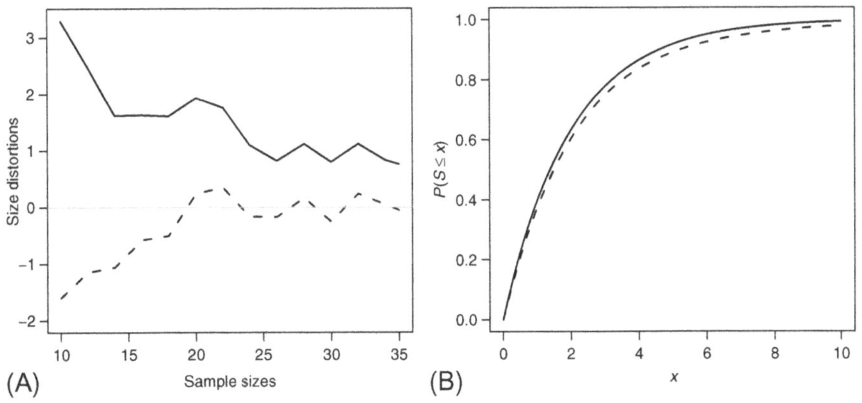

Fig. 3.2 (A) Size distortions of the gradient test (solid) and Bartlett-corrected gradient test (dashes); (B) first-order approximation (solid) and expansion to order $O(n^{-1})$ (dashes) of the null cumulative distribution function of the gradient statistic; gamma distribution.

where $d_1 = 1/[\phi(1-\phi\psi')]$, $d_2 = (1+\phi^2\psi'')/[\phi(1-\phi\psi')^2]$, $d_3 = (2-\phi^3\psi''')/[\phi(1-\phi\psi')^2]$, $d_4 = (1+\phi^2\psi'')^2/[\phi(1-\phi\psi')^3]$, $\psi = \psi(\phi) = \Gamma'(\phi)/\Gamma(\phi)$, $\Gamma'(\phi) = d\Gamma(\phi)/d\phi$, $\psi' = d\psi/d\phi$, $\psi'' = d^2\psi/d\phi^2$, $\psi''' = d^3\psi/d\phi^3$, and $\Gamma(\cdot)$ represents the gamma function.

We now present results from a simulation study with 10,000 replications. The null hypothesis is $\mathcal{H}_0 : (\eta, \phi) = (1, 1)$ in Example (3.5). Note that $\mathcal{H}_0$ means that the data come from an exponential distribution with unity mean. Fig. 3.2A displays a plot of size distortions for the 10% nominal level of the gradient test and its Bartlett-corrected version for different sample sizes. It can be noticed that the gradient test is oversized and its corrected version is clearly less size distorted.

Finally, we set $n = 10$ and plot the first-order and the second-order approximations for the cumulative distribution function of the gradient statistic. Visual inspection of Fig. 3.2B reveals that the first-order χ^2_2 approximation can be inaccurate in small samples.

3.5 BIRNBAUM-SAUNDERS REGRESSION MODELS

3.5.1 The sinh-Normal Distribution

The sinh-normal (SN) distribution with shape, location, and scale parameters given by $\phi > 0$, $\mu \in \mathbb{R}$, and $\sigma > 0$, respectively, was introduced in Rieck and Nedelman [24]. The cumulative distribution function of

$Y \sim \mathrm{SN}(\phi, \mu, \sigma)$ is $F(y; \phi, \mu, \sigma) = \Phi(2\phi^{-1}\sinh[(y-\mu)/\sigma])$, $y \in \mathbb{R}$. The SN probability density function takes the form

$$f(y; \phi, \mu, \sigma) = \frac{2}{\phi\sigma\sqrt{2\pi}} \cosh\left(\frac{y-\mu}{\sigma}\right) \exp\left[-\frac{2}{\phi^2}\sinh^2\left(\frac{y-\mu}{\sigma}\right)\right],$$

where $y \in \mathbb{R}$. The moments of Y can be obtained from the moment generating function

$$M(k) = \exp(k\mu)\left[\frac{B_{k+1/2}(\phi^{-2}) + B_{k-1/2}(\phi^{-2})}{2B_{1/2}(\phi^{-2})}\right],$$

where $B_\nu(\cdot)$ is defined in Example (1.7) and denotes the modified Bessel function of third kind and order ν. The reliability and hazard rate functions of Y are given, respectively, by

$$R(y) = \Phi\left[-\frac{2}{\phi}\sinh\left(\frac{y-\mu}{\sigma}\right)\right], \quad y \in \mathbb{R},$$

$$h(y) = \frac{2\cosh[(y-\mu)/\sigma]\exp\left\{-2\phi^{-2}\sinh^2[(y-\mu)/\sigma]\right\}}{\phi\sigma\sqrt{2\pi}\,\Phi(-2\phi^{-1}\sinh[(y-\mu)/\sigma])}, \quad y \in \mathbb{R}.$$

The SN distribution is symmetrical, presents greater and smaller degrees of kurtosis than the normal model and also has bi-modality. It is symmetric around the mean $\mathbb{E}(Y) = \mu$; it is unimodal for $\phi \leq 2$ and its kurtosis is smaller than that of the normal case; it is bimodal for $\phi > 2$ and its kurtosis is greater than that of the normal case; and if $Y_\phi \sim \mathrm{SN}(\phi, \mu, \sigma)$, then $Z_\phi = 2(Y_\phi - \mu)/(\phi\sigma)$ converges in distribution to the standard normal distribution when $\phi \to 0$.

Rieck and Nedelman [24] proved that if $T \sim \mathrm{BS}(\alpha, \eta)$ (see Section 1.5), then $Y = \log T$ is SN distributed with shape, location, and scale parameters given by $\phi = \alpha$, $\mu = \log \eta$, and $\sigma = 2$, respectively; that is, if $Y \sim \mathrm{SN}(\alpha, \mu, 2)$, then $T = \exp(Y)$ follows the BS distribution with shape parameter α, and scale parameter $\eta = \exp(\mu)$. For this reason, the SN distribution is also called the log-BS distribution. Additionally, the SN and BS models correspond to a logarithmic distribution and its associated distribution, respectively. Fig. 3.3 displays some plots of the SN probability density function for selected values of α with $\mu = 0$ and $\sigma = 2$.

3.5.2 The BS Regression Model

The BS regression model introduced in Rieck and Nedelman [24] is given by

$$y_l = \boldsymbol{x}_l^\top \boldsymbol{\beta} + \varepsilon_l, \quad l = 1, \ldots, n, \tag{3.8}$$

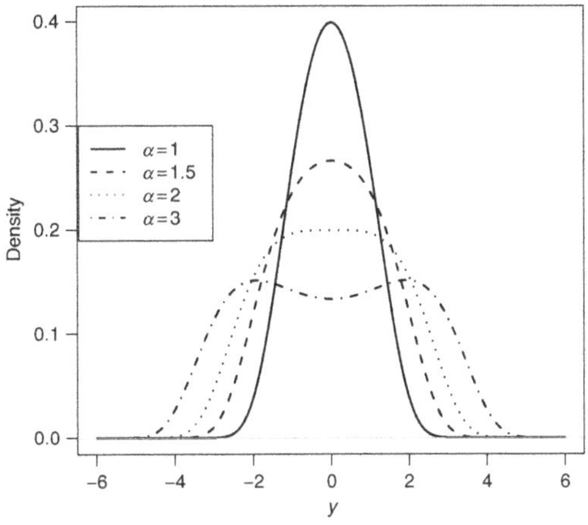

Fig. 3.3 Plots of the SN probability density function: $\mu = 0$ and $\sigma = 2$.

where y_l is the logarithm of the lth observed lifetime, $\boldsymbol{x}_l = (x_{l1}, \ldots, x_{lp})^\top$ is an $p \times 1$ vector of known explanatory variables associated with the lth observable response y_l, and $\boldsymbol{\beta} = (\beta_1, \ldots, \beta_p)^\top$ is a vector of unknown parameters $(p < n)$ to be estimated from the data. The random variables ε_ls are mutually independent errors with SN distribution, that is, $\varepsilon_l \sim \mathrm{SN}(\alpha, 0, 2)$ for $l = 1, \ldots, n$.

The log-likelihood function for the parameter vector $\boldsymbol{\theta} = (\boldsymbol{\beta}^\top, \alpha)^\top$ from a sample $\boldsymbol{y} = (y_1, \ldots, y_n)^\top$ obtained from model (3.8), except for constants, can be expressed as

$$\ell(\boldsymbol{\theta}) = \sum_{l=1}^{n} \log \xi_{l1} - \frac{1}{2} \sum_{l=1}^{n} \xi_{l2}^2,$$

where

$$\xi_{l1} = \xi_{l1}(\boldsymbol{\theta}) = \frac{2}{\alpha} \cosh\left(\frac{y_l - \mu_l}{2}\right), \quad \xi_{l2} = \xi_{l2}(\boldsymbol{\theta}) = \frac{2}{\alpha} \sinh\left(\frac{y_l - \mu_l}{2}\right),$$

and $\mu_l = \boldsymbol{x}_l^\top \boldsymbol{\beta}$. It is assumed that the design matrix $\boldsymbol{X} = (\boldsymbol{x}_1, \ldots, \boldsymbol{x}_n)^\top$ has full column rank. We also assume that some standard regularity conditions on $\ell(\boldsymbol{\theta})$ and its first four derivatives hold as n goes to infinity.

Let $\hat{\boldsymbol{\theta}} = (\hat{\boldsymbol{\beta}}^\top, \hat{\alpha})^\top$ be the MLE of $\boldsymbol{\theta} = (\boldsymbol{\beta}^\top, \alpha)^\top$. A joint iterative algorithm to calculate $\hat{\boldsymbol{\theta}} = (\hat{\boldsymbol{\beta}}^\top, \hat{\alpha})^\top$ is

$$\boldsymbol{X}^{(m)\top} \boldsymbol{X}^{(m)} \boldsymbol{\beta}^{(m+1)} = \boldsymbol{X}^{(m)\top} \boldsymbol{\zeta}^{(m)}, \quad \alpha^{(m+1)} = \frac{\alpha^{(m)}}{2} \left(1 + \bar{\xi}_2^{(m)}\right),$$

where $m = 0, 1, \ldots$ (the iteration counter), $\boldsymbol{\zeta}^{(m)} = \boldsymbol{X}^{(m)}\boldsymbol{\beta}^{(m)} + [4/\psi_1(\alpha^{(m)})]\boldsymbol{s}^{(m)}$, $\boldsymbol{s} = \boldsymbol{s}(\boldsymbol{\theta}) = (s_1, \ldots, s_n)^\top$ with $s_l = (\xi_{l1}\xi_{l2} - \xi_{l2}/\xi_{l1})/2$, and $\bar{\xi}_2^{(m)} = n^{-1}\sum_{l=1}^{n} \xi_{l2}^{2(m)}$. Also,

$$\psi_0(\alpha) = \left[1 - \mathtt{erf}\left(\frac{\sqrt{2}}{\alpha}\right)\right]\exp\left(\frac{2}{\alpha^2}\right), \quad \psi_1(\alpha) = 2 + \frac{4}{\alpha^2} - \frac{\sqrt{2\pi}}{\alpha}\psi_0(\alpha),$$

where $\mathtt{erf}(\cdot)$ is the error function given by $\mathtt{erf}(x) = (2/\sqrt{\pi})\int_0^x \mathrm{e}^{-t^2}\mathrm{d}t$. We can write $\hat{\boldsymbol{\theta}} \stackrel{a}{\sim} \mathcal{N}_{p+1}(\boldsymbol{\theta}, \boldsymbol{K}(\boldsymbol{\theta})^{-1})$ for n large, where $\boldsymbol{K}(\boldsymbol{\theta})$ is the block-diagonal Fisher information matrix given by $\boldsymbol{K}(\boldsymbol{\theta}) = \mathrm{diag}\{\boldsymbol{K}_{\boldsymbol{\beta}}, K_\alpha\}$, $\boldsymbol{K}(\boldsymbol{\theta})^{-1}$ is its inverse, $\boldsymbol{K}_{\boldsymbol{\beta}} = \psi_1(\alpha)(\boldsymbol{X}^\top\boldsymbol{X})/4$ is the information matrix for $\boldsymbol{\beta}$, and $K_\alpha = 2n/\alpha^2$ is the information for α. Since $\boldsymbol{K}(\boldsymbol{\theta})$ is block-diagonal, the vector $\boldsymbol{\beta}$ and the scalar α are globally orthogonal, and $\hat{\boldsymbol{\beta}}$ and $\hat{\alpha}$ are asymptotically independent.

Next, suppose the interest lies in testing the null hypothesis $\mathcal{H}_0 : \boldsymbol{\beta}_1 = \boldsymbol{\beta}_{10}$, which will be tested against the alternative hypothesis $\mathcal{H}_\mathrm{a} : \boldsymbol{\beta}_1 \neq \boldsymbol{\beta}_{10}$, where $\boldsymbol{\beta}$ is partitioned as $\boldsymbol{\beta} = (\boldsymbol{\beta}_1^\top, \boldsymbol{\beta}_2^\top)^\top$, $\boldsymbol{\beta}_1 = (\beta_1, \ldots, \beta_q)^\top$ and $\boldsymbol{\beta}_2 = (\beta_{q+1}, \ldots, \beta_p)^\top$. Here, $\boldsymbol{\beta}_{10}$ is a fixed column vector of dimension q, and $\boldsymbol{\beta}_2$ and α act as nuisance parameters. Let $\tilde{\boldsymbol{\theta}} = (\tilde{\boldsymbol{\beta}}^\top, \tilde{\alpha})^\top$, with $\tilde{\boldsymbol{\beta}} = (\boldsymbol{\beta}_{10}^\top, \tilde{\boldsymbol{\beta}}_2^\top)^\top$, be the restricted MLE of $\boldsymbol{\theta} = (\boldsymbol{\beta}^\top, \alpha)^\top$. The gradient statistic for testing $\mathcal{H}_0$ takes the form

$$S_\mathrm{T} = \tilde{\boldsymbol{s}}^\top \boldsymbol{X}_1(\hat{\boldsymbol{\beta}}_1 - \boldsymbol{\beta}_{10}),$$

where the matrix $\boldsymbol{X}$ is partitioned as $\boldsymbol{X} = [\boldsymbol{X}_1 \;\; \boldsymbol{X}_2]$, $\boldsymbol{X}_1$ being $n \times q$ and $\boldsymbol{X}_2$ being $n \times (p - q)$, and $\tilde{\boldsymbol{s}} = \boldsymbol{s}(\tilde{\boldsymbol{\theta}})$. The limiting distribution of this statistic under $\mathcal{H}_0$ is χ_q^2.

3.5.3 The Bartlett-Corrected Gradient Statistic

To define the corrected gradient statistic, some additional notation is in order. We define the matrices

$$\begin{aligned}
\boldsymbol{Z} &= \boldsymbol{X}(\boldsymbol{X}^\top\boldsymbol{X})^{-1}\boldsymbol{X}^\top = ((z_{lc})), \\
\boldsymbol{Z}_2 &= \boldsymbol{X}_2(\boldsymbol{X}_2^\top\boldsymbol{X}_2)^{-1}\boldsymbol{X}_2^\top = ((z_{2lc})), \\
\boldsymbol{Z}_d &= \mathrm{diag}\{z_{11}, \ldots, z_{nn}\}, \quad \boldsymbol{Z}_{2d} = \mathrm{diag}\{z_{211}, \ldots, z_{2nn}\}.
\end{aligned}$$

Additionally, we define the quantities

$$g_1(\alpha) = \frac{96\psi_2(\alpha)}{\psi_1(\alpha)^2}, \quad g_2(\alpha) = -\frac{48\psi_2(\alpha)}{\psi_1(\alpha)^2}, \quad g_3(\alpha) = \frac{60(2 + \alpha^2)}{\alpha^2\psi_1(\alpha)},$$

$$g_4(\alpha) = \frac{4(12+\alpha^2)}{\psi_1(\alpha)}, \quad g_5(\alpha) = -\frac{96(2+\alpha^2)\psi_3(\alpha)}{\alpha\psi_1(\alpha)^2},$$
$$g_6(\alpha) = \frac{48(2+\alpha^2)^2}{\alpha^4\psi_1(\alpha)^2}, \quad g_7(\alpha) = -\frac{24(2+\alpha^2)^2}{\alpha^4\psi_1(\alpha)^2},$$
$$\psi_2(\alpha) = -\frac{1}{4}\left[2+\frac{7}{\alpha^2}-\sqrt{\frac{\pi}{2}}\left(\frac{1}{2\alpha}+\frac{6}{\alpha^3}\right)\psi_0(\alpha)\right],$$
$$\psi_3(\alpha) = \frac{3}{\alpha^3}-\frac{\sqrt{2\pi}}{4\alpha^2}\left(1+\frac{4}{\alpha^2}\right)\psi_0(\alpha).$$

We also define $\boldsymbol{Z}^{(2)} = \boldsymbol{Z} \odot \boldsymbol{Z}$, $\boldsymbol{Z}_d^{(2)} = \boldsymbol{Z}_d \odot \boldsymbol{Z}_d$, $\boldsymbol{Z}_2^{(2)} = \boldsymbol{Z}_2 \odot \boldsymbol{Z}_2$, etc., and "$\odot$" denotes the Hadamard (elementwise) product of matrices.

The Bartlett-corrected gradient statistic is given by $S_{\mathrm{T}}^* = S_{\mathrm{T}}[1-(c+bS_{\mathrm{T}}+aS_{\mathrm{T}}^2)]$, where $a = A_3/[12q(q+2)(q+4)]$, $b = (A_2+A_{2,\beta\alpha}-2A_3)/[12q(q+2)]$, $c = (A_1+A_{1,\beta\alpha}-A_2-A_{2,\beta\alpha}+A_3)/(12q)$,

$$A_1 = g_1(\alpha)\mathrm{tr}\{(\boldsymbol{Z}_d-\boldsymbol{Z}_{2d})\boldsymbol{Z}_{2d}\}, \quad A_2 = g_2(\alpha)\mathrm{tr}\{(\boldsymbol{Z}_d-\boldsymbol{Z}_{2d})^{(2)}\},$$
$$A_{1,\beta\alpha} = \frac{q}{n}[g_3(\alpha)+g_4(\alpha)+g_5(\alpha)+(p-q)g_6(\alpha)],$$
$$A_{2,\beta\alpha} = \frac{q(q+2)g_7(\alpha)}{n},$$

and $A_3 = 0$. If α is known, then $A_{1,\beta\alpha}$ and $A_{2,\beta\alpha}$ are zero. Since $A_3 = 0$, we have that $a = 0$ and hence the Bartlett-corrected gradient statistic takes the form

$$S_{\mathrm{T}}^* = S_{\mathrm{T}}[1-(c+bS_{\mathrm{T}})], \tag{3.9}$$

where

$$b = \frac{A_2+A_{2,\beta\alpha}}{12q(q+2)}, \quad c = \frac{A_1-A_2+A_{1,\beta\alpha}-A_{2,\beta\alpha}}{12q}.$$

The null distribution of S_{T}^* is χ^2 with approximation error reduced from order $O(n^{-1})$ to $O(n^{-3/2})$. Also, all unknown parameters in the quantities that define the Bartlett-corrected gradient statistic are replaced by their restricted MLEs. The improved gradient test that employ Eq. (3.9) as test statistic, follow from the comparison of the observed value of S_{T}^* with the critical value obtained as the appropriate χ_q^2 quantile.

A brief commentary on the quantities that define the improved gradient statistic is in order. Note that A_1 and A_2 depend on the model matrix $\boldsymbol{X}$ in question. They also involve the shape parameter α. Unfortunately, they are not easy to interpret in generality and provide no indication as to what structural aspects of the model contribute significantly to their magnitude. The quantities $A_{1,\beta\alpha}$ and $A_{2,\beta\alpha}$ can be regarded as the contribution due to the fact that the shape parameter α is considered unknown and has to be estimated from the data. Notice that $A_{1,\beta\alpha}$ depends on the model matrix $\boldsymbol{X}$ only through its rank, that is, the number of regression parameters (p), and it also involves the number of parameters of interest (q) in the null hypothesis. Additionally, $A_{2,\beta\alpha}$ involves the number of parameters of interest. Therefore, it implies that these quantities can be non-negligible if the dimension of $\boldsymbol{\beta}$ and/or the number of tested parameters in the null hypothesis are not considerably smaller than the sample size.

3.5.4 Numerical Evidence

We shall now report Monte Carlo evidence on the finite sample performance of the gradient test and its Bartlett-corrected version in BS regressions. The simulations were based on the model

$$y_l = \beta_1 x_{l1} + \beta_2 x_{l2} + \cdots + \beta_p x_{lp} + \varepsilon_l,$$

where $x_{l1} = 1$ and $\varepsilon_l \sim \mathrm{SN}(\alpha, 0, 2)$, $l = 1, \ldots, n$. The covariate values were selected as random draws from the uniform $\mathcal{U}(0, 1)$ distribution and, for fixed n, those values were kept constant throughout the experiment. The number of Monte Carlo replications was 15,000, the nominal levels of the tests were $\gamma = 10\%$ and 5%, and all simulations were performed using the `Ox` matrix programming language (http://www.doornik.com).

Table 3.1 lists the null rejection rates (entries are percentages) of the two tests. The null hypothesis is $\mathcal{H}_0 : \beta_1 = \beta_2 = 0$, which is tested against a two-sided alternative, the sample size are $n = 20, 40, 60$, and 100, and $\alpha = 0.5$. Different values of p were considered. The values of the response were generated using $\beta_3 = \beta_4 = \cdots = \beta_p = 1$. Note that the usual gradient test is oversized (liberal), more so as the number of regressors increases. On the other hand, the Bartlett-corrected gradient test that uses S_{T}^* as test statistic is less size distorted than the usual gradient test that employs S_{T} as test statistic. Note that the improved gradient test produced null rejection rates that are close to the nominal levels in all cases considered. The values in this table show that the null rejection rates of the two tests approach the corresponding nominal levels as the sample size grows, as expected.

Table 3.1 Null Rejection Rates (%) for $\mathcal{H}_0: \beta_1 = \beta_2 = 0$; $\alpha = 0.5$ and Different Sample Sizes

	$n = 20$				$n = 40$			
	S_T		S_T^*		S_T		S_T^*	
p	10%	5%	10%	5%	10%	5%	10%	5%
3	10.73	4.91	8.89	4.28	10.13	5.05	9.32	4.74
4	12.13	5.65	8.81	4.14	11.27	5.55	9.60	4.75
5	14.11	7.10	9.39	4.59	11.91	5.66	9.69	4.43
6	15.65	7.95	9.11	4.57	12.18	6.18	9.45	4.52
	$n = 60$				$n = 100$			
	S_T		S_T^*		S_T		S_T^*	
p	10%	5%	10%	5%	10%	5%	10%	5%
3	10.17	5.11	9.58	4.94	10.23	5.21	9.75	5.09
4	10.71	5.08	9.74	4.71	10.47	5.17	9.83	4.93
5	11.67	5.78	10.25	5.01	11.01	5.31	10.11	4.82
6	11.95	6.01	10.12	4.97	11.09	5.36	9.93	4.77

3.6 GENERALIZED LINEAR MODELS

Suppose the univariate variables $y_1, \ldots, y_n$ are independent and each y_l has a probability density function in the following family of distributions:

$$f(y; \theta_l, \phi) = \exp\{\phi[y\theta_l - b(\theta_l) + c(y)] + a(y, \phi)\}, \quad l = 1, \ldots, n, \tag{3.10}$$

where $a(\cdot,\cdot)$, $b(\cdot)$, and $c(\cdot)$ are known appropriate functions. The mean and variance are $\mathbb{E}(y_l) = \mu_l = db(\theta_l)/d\theta_l$ and $\mathbb{VAR}(y_l) = \phi_l^{-1} V_l$, where $V_l = d\mu_l/d\theta_l$ is called the variance function and $\theta_l = q(\mu_l) = \int V_l^{-1}\, d\mu_l$ is a known one-to-one function of μ_l. In order to introduce a regression structure in the class of models in Eq. (3.10), we assume that

$$d(\mu_l) = \eta_l = \boldsymbol{x}_l^\top \boldsymbol{\beta}, \quad l = 1, \ldots, n, \tag{3.11}$$

where $d(\cdot)$ is a known one-to-one differentiable link function, $\boldsymbol{x}_l = (x_{l1}, \ldots, x_{lp})^\top$ is a vector of known variables associated with the lth observable response, and $\boldsymbol{\beta} = (\beta_1, \ldots, \beta_p)^\top$ is a set of unknown parameters to be estimated from the data ($p < n$). The regression structure links the covariates $\boldsymbol{x}_l$ to the parameter of interest μ_l. Here, we assume only identifiability in the sense that distinct βs imply distinct ηs. Further, the precision parameter may be known or unknown, and it is the same for all observations.

Let $\ell(\boldsymbol{\beta}, \phi)$ denote the total log-likelihood function for a given generalized linear model (GLM). We have

$$\ell(\boldsymbol{\beta}, \phi) = \phi \sum_{l=1}^{n} [y_l\theta_l - b(\theta_l) + c(y_l)] + \sum_{l=1}^{n} a(y_l, \phi),$$

where θ_l is related to $\boldsymbol{\beta}$ through Eq. (3.11) as $d(db(\theta_l)/d\theta_l) = \boldsymbol{x}_l^\top \boldsymbol{\beta}$. The score function and Fisher information matrix for $\boldsymbol{\beta}$ are given, respectively, by $\boldsymbol{U}_{\boldsymbol{\beta}}(\boldsymbol{\beta}, \phi) = \phi \boldsymbol{X}^\top \boldsymbol{W}^{1/2} \boldsymbol{V}^{-1/2}(\boldsymbol{y} - \boldsymbol{\mu})$ and $\boldsymbol{K}_{\boldsymbol{\beta}} = \phi \boldsymbol{X}^\top \boldsymbol{W} \boldsymbol{X}$, where $\boldsymbol{W} = \mathrm{diag}\{w_1, \ldots, w_n\}$ with $w_l = V_l^{-1}(d\mu_l/d\eta_l)^2$, $\boldsymbol{V} = \mathrm{diag}\{V_1, \ldots, V_n\}$, $\boldsymbol{y} = (y_1, \ldots, y_n)^\top$ and $\boldsymbol{\mu} = (\mu_1, \ldots, \mu_n)^\top$. The model matrix $\boldsymbol{X} = (\boldsymbol{x}_1, \ldots, \boldsymbol{x}_n)^\top$ is assumed to be of full rank, that is, $\mathrm{rank}(\boldsymbol{X}) = p$. The MLE $\hat{\boldsymbol{\beta}}$ of $\boldsymbol{\beta}$ can be obtained iteratively using the standard reweighted least squares method

$$\boldsymbol{X}^\top \boldsymbol{W}^{(m)} \boldsymbol{X} \boldsymbol{\beta}^{(m+1)} = \boldsymbol{X}^\top \boldsymbol{W}^{(m)} \boldsymbol{y}^{*(m)}, \quad m = 0, 1, \ldots,$$

where $\boldsymbol{y}^{*(m)} = \boldsymbol{X}\boldsymbol{\beta}^{(m)} + \boldsymbol{N}^{(m)}(\boldsymbol{y} - \boldsymbol{\mu}^{(m)})$ is a modified dependent variable, and $\boldsymbol{N} = \mathrm{diag}\{(d\mu_1/d\eta_1)^{-1}, \ldots, (d\mu_n/d\eta_n)^{-1}\}$. The above equation shows that any software with a weighted regression routine can be used to evaluate $\hat{\boldsymbol{\beta}}$. Additionally, note that $\hat{\boldsymbol{\beta}}$ does not depend on the parameter ϕ.

Estimation of the precision parameter ϕ by the maximum likelihood method is a more difficult problem than the estimation of $\boldsymbol{\beta}$ and the complexity depends on the functional form of $a(y, \phi)$. The MLE $\hat{\phi}$ of ϕ is a function of the deviance (D_p) of the model, which is defined as $D_p = 2\sum_{l=1}^{n}[v(y_l) - v(\hat{\mu}_l) + (\hat{\mu}_l - y_l)q(\hat{\mu}_l)]$, where $v(z) = zq(z) - b(q(z))$ and $\hat{\mu}_l$ denotes the MLE of μ_l $(l = 1, \ldots, n)$. That is, given the estimate $\hat{\boldsymbol{\beta}}$, the MLE of ϕ can be found as the solution of the equation

$$\sum_{l=1}^{n} \left. \frac{\partial a(y_l, \phi)}{\partial \phi} \right|_{\phi = \hat{\phi}} = \frac{D_p}{2} - \sum_{l=1}^{n} v(y_l).$$

When Eq. (3.10) is a two-parameter full exponential family distribution with canonical parameters ϕ and $\phi\theta$, the term $a(y, \phi)$ in Eq. (3.10) can be expressed as $a(y, \phi) = \phi a_0(y) + a_1(\phi) + a_2(y)$, and the estimate of ϕ is obtained from

$$a_1'(\hat{\phi}) = \frac{1}{n}\left[\frac{D_p}{2} - \sum_{l=1}^{n} t(y_l)\right],$$

where $a_1'(\phi) = da_1(\phi)/d\phi$ and $t(y_l) = v(y_l) + a_0(y_l)$, for $l = 1, \ldots, n$. Table 3.2 lists the functions $a_1(\phi)$, $v(y)$, and $t(y)$ for normal, inverse Gaussian and gamma models. For normal and inverse Gaussian models we

Table 3.2 Some Special Models

Model	$a_1(\phi)$	$v(y)$	$t(y)$
Normal	$(1/2)\log\phi$	$y^2/2$	0
Inverse Gaussian	$(1/2)\log\phi$	$1/(2y)$	0
Gamma	$\phi\log\phi - \log\Gamma(\phi)$	$\log y - 1$	-1

have that $\hat{\phi} = n/D_p$, whereas for the gamma model the MLE $\hat{\phi}$ is obtained from $\log\hat{\phi} - \psi(\hat{\phi}) = D_p/(2n)$, where $\psi(\cdot)$ denotes the digamma function, thus requiring the use of a nonlinear numerical algorithm.

Next, we shall consider the test that is based on the gradient statistic in the class of GLMs for testing a composite null hypothesis. The hypothesis of interest is $\mathcal{H}_0 : \boldsymbol{\beta}_1 = \boldsymbol{\beta}_{10}$, which will be tested against the alternative hypothesis $\mathcal{H}_a : \boldsymbol{\beta}_1 \neq \boldsymbol{\beta}_{10}$, where $\boldsymbol{\beta}$ is partitioned as $\boldsymbol{\beta} = (\boldsymbol{\beta}_1^\top, \boldsymbol{\beta}_2^\top)^\top$, $\boldsymbol{\beta}_1 = (\beta_1, \ldots, \beta_q)^\top$, and $\boldsymbol{\beta}_2 = (\beta_{q+1}, \ldots, \beta_p)^\top$. Here, $\boldsymbol{\beta}_{10}$ is a fixed column vector of dimension q, and $\boldsymbol{\beta}_2$ and ϕ act as nuisance parameters. Let $(\hat{\boldsymbol{\beta}}_1, \hat{\boldsymbol{\beta}}_2, \hat{\phi})$ and $(\boldsymbol{\beta}_{10}, \tilde{\boldsymbol{\beta}}_2, \tilde{\phi})$ be the unrestricted and restricted MLEs of $(\boldsymbol{\beta}_1, \boldsymbol{\beta}_2, \phi)$, respectively. The gradient statistic for testing $\mathcal{H}_0$ can be expressed as

$$S_T = \tilde{\phi}^{1/2}\tilde{\boldsymbol{s}}^\top \tilde{\boldsymbol{W}}^{1/2}\boldsymbol{X}_1(\hat{\boldsymbol{\beta}}_1 - \boldsymbol{\beta}_{10}),$$

where the matrix $\boldsymbol{X}$ is partitioned as $\boldsymbol{X} = [\boldsymbol{X}_1 \ \boldsymbol{X}_2]$, $\boldsymbol{X}_1$ being $n \times q$ and $\boldsymbol{X}_2$ being $n \times (p - q)$, and $\boldsymbol{s} = \phi^{1/2}\boldsymbol{V}^{-1/2}(\boldsymbol{y} - \boldsymbol{\mu})$ is the Pearson residual vector. Here, tildes indicate quantities available at the restricted MLEs. Under the null hypothesis, this statistic has a χ_q^2 distribution up to an error of order $O(n^{-1})$.

3.6.1 The Bartlett-Corrected Gradient Statistic

To define the modified gradient statistic in GLMs, some additional notation is in order. We define the matrices $\boldsymbol{Z} = \boldsymbol{X}(\boldsymbol{X}^\top\boldsymbol{W}\boldsymbol{X})^{-1}\boldsymbol{X}^\top = ((z_{lc}))$, $\boldsymbol{Z}_2 = \boldsymbol{X}_2(\boldsymbol{X}_2^\top\boldsymbol{W}\boldsymbol{X}_2)^{-1}\boldsymbol{X}_2^\top = ((z_{2lc}))$, $\boldsymbol{Z}_d = \text{diag}\{z_{11}, \ldots, z_{nn}\}$, $\boldsymbol{Z}_{2d} = \text{diag}\{z_{211}, \ldots, z_{2nn}\}$, $\boldsymbol{F} = \text{diag}\{f_1, \ldots, f_n\}$, $\boldsymbol{G} = \text{diag}\{g_1, \ldots, g_n\}$, $\boldsymbol{T} = \text{diag}\{t_1, \ldots, t_n\}$, $\boldsymbol{D} = \text{diag}\{d_1, \ldots, d_n\}$, and $\boldsymbol{E} = \text{diag}\{e_1, \ldots, e_n\}$, where

$$f_l = \frac{1}{V_l}\frac{d\mu_l}{d\eta_l}\frac{d^2\mu_l}{d\eta_l^2}, \quad g_l = f_l - \frac{1}{V_l^2}\frac{dV_l}{d\mu_l}\left(\frac{d\mu_l}{d\eta_l}\right)^3,$$

$$\lambda_{1l} = \frac{1}{V_l^2}\frac{dV_l}{d\mu_l}\left(\frac{d\mu_l}{d\eta_l}\right)^2\frac{d^2\mu_l}{d\eta_l^2}, \quad \lambda_{2l} = \frac{1}{V_l}\left(\frac{d^2\mu_l}{d\eta_l^2}\right)^2,$$

$$\lambda_{3l} = \frac{1}{V_l}\frac{d\mu_l}{d\eta_l}\frac{d^3\mu_l}{d\eta_l^3}, \quad \lambda_{4l} = \frac{1}{V_l^3}\left(\frac{dV_l}{d\mu_l}\right)^2\left(\frac{d\mu_l}{d\eta_l}\right)^4,$$

$$\lambda_{5l} = \frac{1}{V_l^2}\frac{d^2V_l}{d\mu_l^2}\left(\frac{d\mu_l}{d\eta_l}\right)^4, \quad t_l = -9\lambda_{1l} + 3\lambda_{2l} + 3\lambda_{3l} + 4\lambda_{4l} - 2\lambda_{5l},$$

$$d_l = -5\lambda_{1l} + 2\lambda_{2l} + 2\lambda_{3l} + 2\lambda_{4l} - \lambda_{5l},$$

$$e_l = -12\lambda_{1l} + 3\lambda_{2l} + 4\lambda_{3l} + 6\lambda_{4l} - 3\lambda_{5l}.$$

We also define $\boldsymbol{Z}^{(2)} = \boldsymbol{Z} \odot \boldsymbol{Z}$, $\boldsymbol{Z}_2^{(2)} = \boldsymbol{Z}_2 \odot \boldsymbol{Z}_2$, $\boldsymbol{Z}^{(3)} = \boldsymbol{Z}^{(2)} \odot \boldsymbol{Z}$, etc. The matrices $\phi^{-1}\boldsymbol{Z}$ and $\phi^{-1}\boldsymbol{Z}_2$ have simple interpretations as asymptotic covariance structures of $\boldsymbol{X}\hat{\boldsymbol{\beta}}$ and $\boldsymbol{X}_2\tilde{\boldsymbol{\beta}}_2$, respectively. Let $\mathbf{1}_n = (1, \ldots, 1)^\top$ be the n-vector of ones. Also, $d_{(2)} = d_{(2)}(\phi) = \phi^2 a_1''(\phi)$ and $d_{(3)} = d_{(3)}(\phi) = \phi^3 a_1'''(\phi)$, where $a_1''(\phi) = da_1'(\phi)/d\phi$ and $a_1'''(\phi) = da_1''(\phi)/d\phi$.

The Bartlett-corrected gradient statistic for testing $\mathcal{H}_0 : \boldsymbol{\beta}_1 = \boldsymbol{\beta}_{10}$ in GLMs was derived in Vargas et al. [25], and it is given by

$$S_{\mathrm{T}}^* = S_{\mathrm{T}}\left[1 - \left(c + bS_{\mathrm{T}} + aS_{\mathrm{T}}^2\right)\right], \tag{3.12}$$

where $a = A_3/[12q(q+2)(q+4)]$, $b = (A_2 + A_{2,\beta\phi} - 2A_3)/[12q(q+2)]$, $c = (A_1 + A_{1,\beta\phi} - A_2 - A_{2,\beta\phi} + A_3)/(12q)$,

$$\begin{aligned}
A_1 = {} & 12\phi^{-1}\mathbf{1}_n^\top(\boldsymbol{F}+\boldsymbol{G})[\boldsymbol{Z}_d\boldsymbol{Z}\boldsymbol{Z}_d - \boldsymbol{Z}_{2d}\boldsymbol{Z}_2\boldsymbol{Z}_{2d} \\
& \quad + \boldsymbol{Z}^{(3)} - \boldsymbol{Z}_2^{(3)}](\boldsymbol{F}+\boldsymbol{G})\mathbf{1}_n \\
& - 6\phi^{-1}\mathbf{1}_n^\top(\boldsymbol{F}+2\boldsymbol{G})[(\boldsymbol{Z}+\boldsymbol{Z}_2)\odot(\boldsymbol{Z}^{(2)} - \boldsymbol{Z}_2^{(2)}) \\
& \quad + 2\boldsymbol{Z}_{2d}(\boldsymbol{Z}\boldsymbol{Z}_d - \boldsymbol{Z}_2\boldsymbol{Z}_{2d}) + 2\boldsymbol{Z}_2^{(2)}\odot(\boldsymbol{Z}-\boldsymbol{Z}_2) \\
& \quad + (\boldsymbol{Z}-\boldsymbol{Z}_2)_d(\boldsymbol{Z}\boldsymbol{Z}_d + \boldsymbol{Z}_2\boldsymbol{Z}_{2d})](\boldsymbol{F}+\boldsymbol{G})\mathbf{1}_n \\
& + 3\phi^{-1}\mathbf{1}_n^\top(\boldsymbol{F}+2\boldsymbol{G})[2(\boldsymbol{Z}-\boldsymbol{Z}_2)_d\boldsymbol{Z}_2\boldsymbol{Z}_{2d} + 2\boldsymbol{Z}_2^{(2)}\odot(\boldsymbol{Z}-\boldsymbol{Z}_2) \\
& \quad + \boldsymbol{Z}_{2d}(\boldsymbol{Z}-\boldsymbol{Z}_2)(\boldsymbol{Z}-\boldsymbol{Z}_2)_d + \boldsymbol{Z}_{2d}(\boldsymbol{Z}-\boldsymbol{Z}_2)\boldsymbol{Z}_{2d}](\boldsymbol{F}+2\boldsymbol{G})\mathbf{1}_n \\
& + 6\phi^{-1}\mathbf{1}_n^\top\boldsymbol{T}(\boldsymbol{Z}-\boldsymbol{Z}_2)_d(\boldsymbol{Z}_d + 3\boldsymbol{Z}_{2d})\mathbf{1}_n \\
& - 12\phi^{-1}\mathbf{1}_n^\top\boldsymbol{D}(\boldsymbol{Z}_d^{(2)} - \boldsymbol{Z}_{2d}^{(2)})\mathbf{1}_n - 6\phi^{-1}\mathbf{1}_n^\top\boldsymbol{E}(\boldsymbol{Z}-\boldsymbol{Z}_2)_d\boldsymbol{Z}_{2d}\mathbf{1}_n,
\end{aligned}$$

$$
\begin{aligned}
A_2 = &-3\phi^{-1}\mathbf{1}_n^\top(\boldsymbol{F}+2\boldsymbol{G})\left[\frac{3}{4}(\boldsymbol{Z}-\boldsymbol{Z}_2)_d(\boldsymbol{Z}-\boldsymbol{Z}_2)(\boldsymbol{Z}-\boldsymbol{Z}_2)_d\right.\\
&+\frac{1}{2}(\boldsymbol{Z}-\boldsymbol{Z}_2)^{(3)}+\boldsymbol{Z}_{2d}(\boldsymbol{Z}-\boldsymbol{Z}_2)(\boldsymbol{Z}-\boldsymbol{Z}_2)_d\\
&+(\boldsymbol{Z}-\boldsymbol{Z}_2)_d\boldsymbol{Z}_2(\boldsymbol{Z}-\boldsymbol{Z}_2)_d\\
&\left.+2\boldsymbol{Z}_2\odot(\boldsymbol{Z}-\boldsymbol{Z}_2)^{(2)}\right](\boldsymbol{F}+2\boldsymbol{G})\mathbf{1}_n\\
&+6\phi^{-1}\mathbf{1}_n^\top(\boldsymbol{F}+2\boldsymbol{G})[(\boldsymbol{Z}-\boldsymbol{Z}_2)\odot(\boldsymbol{Z}^{(2)}-\boldsymbol{Z}_2^{(2)})\\
&+(\boldsymbol{Z}-\boldsymbol{Z}_2)_d(\boldsymbol{Z}\boldsymbol{Z}_d-\boldsymbol{Z}_2\boldsymbol{Z}_{2d})](\boldsymbol{F}+\boldsymbol{G})\mathbf{1}_n\\
&-3\phi^{-1}\mathbf{1}_n^\top(2\boldsymbol{T}-\boldsymbol{E})(\boldsymbol{Z}-\boldsymbol{Z}_2)_d^{(2)}\mathbf{1}_n,\\
A_3 = &\,\phi^{-1}\mathbf{1}_n^\top(\boldsymbol{F}+2\boldsymbol{G})\left[\frac{3}{4}(\boldsymbol{Z}-\boldsymbol{Z}_2)_d(\boldsymbol{Z}-\boldsymbol{Z}_2)(\boldsymbol{Z}-\boldsymbol{Z}_2)_d\right.\\
&\left.+\frac{1}{2}(\boldsymbol{Z}-\boldsymbol{Z}_2)^{(3)}\right](\boldsymbol{F}+2\boldsymbol{G})\mathbf{1}_n,\\
A_{1,\beta\phi} = &\,\frac{6q[d_{(3)}+(2-p+q)d_{(2)}]}{nd_{(2)}^2},\quad A_{2,\beta\phi}=\frac{3q(q+2)}{nd_{(2)}};
\end{aligned}
$$

when ϕ is known $A_{1,\beta\phi}$ and $A_{2,\beta\phi}$ are zero. The notation $(\cdot)_d$ indicates that the off-diagonal elements of the matrix were set equal to zero. The modified statistic S_T^* has a χ_q^2 distribution up to an error of order $O(n^{-3/2})$ under the null hypothesis.

Note that A_1, A_2, and A_3 depend heavily on the particular model matrix $\boldsymbol{X}$ in question. They involve the (possibly unknown) dispersion parameter and the unknown means. Further, they depend on the mean link function and its first, second, and third derivatives. They also involve the variance function and its first and second derivatives. Unfortunately, they are not easy to interpret in generality and provide no indication as to what structural aspects of the model contribute significantly to their magnitude. The quantities $A_{1,\beta\phi}$ and $A_{2,\beta\phi}$ can be regarded as the contribution yielded by the fact that ϕ is considered unknown and has to be estimated from the data. Notice that $A_{1,\beta\phi}$ depends on the model matrix only through its rank, that is, the number of regression parameters (p), and it also involves the number of parameters of interest (q) in the null hypothesis. Additionally, $A_{2,\beta\phi}$ involves the number of parameters of interest. Therefore, it implies that these quantities can be non-negligible if the dimension of $\boldsymbol{\beta}$ and/or the number of tested parameters in the null hypothesis are not considerably smaller than the sample size.

Notice that the general expressions which define the improved gradient statistic only involve simple operations on matrices and vectors, and hence can be easily implemented in any mathematical or statistical/econometric programming environment. Also, all unknown parameters in the quantities that define the Bratlett-corrected gradient statistic are replaced by their restricted MLEs.

3.6.2 Finite-Sample Performance

We report the results from Monte Carlo simulation experiments in order to compare the performance of the usual gradient (S_{T}) test and the improved gradient (S_{T}^{*}) test in small- and moderate-sized samples for testing hypotheses in the class of GLMs. We assume that

$$d(\mu_l) = \log \mu_l = \eta_l = \beta_1 x_{l1} + \beta_2 x_{l2} + \cdots + \beta_p x_{lp}, \quad l = 1, \ldots, n,$$

where $\phi > 0$ is assumed unknown and it is the same for all observations. The number of Monte Carlo replications was 15,000, and the nominal levels of the tests were $\gamma = 10\%, 5\%$, and 1%. The simulations were carried out using the Ox matrix programming language (http://www.doornik.com). All regression parameters, except those fixed at the null hypothesis, were set equal to one. The simulation results are based on the gamma and inverse Gaussian regression models. For the gamma model, we set $\phi = 1$ and the covariate values were selected as random draws from the $\mathcal{U}(0,1)$ distribution. We set $\phi = 3$ and selected the covariate values as random draws from the $\mathcal{N}(0,1)$ distribution for the inverse Gaussian model. For each fixed n, the covariate values were kept constant throughout the experiment for both gamma and inverse Gaussian regression models.

We report the null rejection rates of $\mathcal{H}_0 : \beta_1 = \cdots = \beta_q = 0$ for the tests at the 10%, 5%, and 1% nominal significance levels, that is, the percentage of times that the corresponding statistics exceed the 10%, 5%, and 1% upper points of the reference χ^2 distribution. The results are presented in Tables 3.3 and 3.4 for the gamma model, whereas Tables 3.5 and 3.6 report the results for the inverse Gaussian model. Entries are percentages. We consider different values for p (number of regression parameters), q (number of tested parameters in the null hypothesis), and n (sample size).

The values in Tables 3.3–3.6 reveal important information. The usual gradient test can be oversized (liberal) to test hypotheses on the model parameters in GLMs, rejecting the null hypothesis more frequently than expected based on the selected nominal levels. On the other hand, the improved gradient test that employs S_{T}^{*} as test statistic is less size distorted

Table 3.3 Null Rejection Rates (%) for $\mathcal{H}_0 : \beta_1 = \cdots = \beta_q = 0$ With $p = 4$; Gamma Model

q	n	γ(%)	S_T	S_T^*	q	n	γ(%)	S_T	S_T^*
1	20	10	13.02	10.39	2	20	10	11.71	10.07
		5	6.68	5.23			5	5.61	5.01
		1	1.15	0.96			1	0.73	0.91
	25	10	12.38	10.39		25	10	11.69	10.07
		5	6.35	5.20			5	5.59	5.07
		1	1.19	1.11			1	0.63	0.76
	30	10	11.61	9.89		30	10	11.46	10.31
		5	5.97	5.04			5	5.30	4.94
		1	0.99	0.95			1	0.76	0.90

Table 3.4 Null Rejection Rates (%) for $\mathcal{H}_0 : \beta_1 = \cdots = \beta_q = 0$ With $p = 6$; Gamma Model

q	n	γ(%)	S_T	S_T^*	q	n	γ(%)	S_T	S_T^*
2	20	10	15.24	10.27	3	20	10	13.57	10.14
		5	7.57	4.88			5	6.18	4.92
		1	1.10	0.85			1	0.61	0.70
	25	10	14.03	10.4		25	10	12.51	9.95
		5	7.03	4.97			5	5.75	4.77
		1	1.04	0.84			1	0.76	0.81
	30	10	12.99	10.01		30	10	11.87	9.65
		5	6.37	5.04			5	5.69	4.70
		1	0.97	0.82			1	0.67	0.69

Table 3.5 Null Rejection Rates (%) for $\mathcal{H}_0 : \beta_1 = \cdots = \beta_q = 0$ With $p = 4$; Inverse Gaussian Model

q	n	γ(%)	S_T	S_T^*	q	n	γ(%)	S_T	S_T^*
1	20	10	13.19	9.94	2	20	10	9.15	9.90
		5	6.46	4.97			5	3.27	4.71
		1	1.11	0.97			1	0.21	0.73
	25	10	12.41	10.63		25	10	10.51	9.95
		5	5.86	5.19			5	4.55	4.93
		1	0.77	0.96			1	0.31	0.71
	30	10	11.87	9.99		30	10	11.03	10.46
		5	5.83	5.14			5	4.76	4.92
		1	0.85	0.84			1	0.49	0.85

Table 3.6 Null Rejection Rates (%) for $\mathcal{H}_0 : \beta_1 = \cdots = \beta_q = 0$ With $p = 6$; Inverse Gaussian Model

q	n	$\gamma(\%)$	S_T	S_T^*	q	n	$\gamma(\%)$	S_T	S_T^*
2	20	10	14.85	10.59	3	20	10	12.83	11.00
		5	7.44	5.18			5	5.47	5.41
		1	0.85	0.83			1	0.51	1.03
	25	10	14.28	10.54		25	10	12.55	10.84
		5	7.09	5.58			5	5.40	5.27
		1	0.91	0.94			1	0.71	1.09
	30	10	11.83	10.07		30	10	10.67	9.78
		5	5.20	4.83			5	4.30	4.67
		1	0.75	1.05			1	0.33	0.71

than the usual gradient test for testing hypotheses in GLMs; that is, the impact of the number of regressors and the number of tested parameters in the null hypothesis are much less important for the improved test. Also, the improved gradient test produced null rejection rates that are very close to the nominal levels in all cases considered. Finally, the values in Tables 3.3–3.6 show that the null rejection rates of the tests approach the corresponding nominal levels as the sample size grows, as expected.

3.6.3 Tests on the Parameter ϕ

In this section, the problem under consideration is that of testing a composite null hypothesis $\mathcal{H}_0 : \phi = \phi_0$ against $\mathcal{H}_a : \phi \neq \phi_0$, where ϕ_0 is a positive specified value for ϕ. Here, $\boldsymbol{\beta}$ acts as a vector of nuisance parameters. The gradient statistic for testing $\mathcal{H}_0 : \phi = \phi_0$ can be expressed as

$$S_T = n[a_1'(\phi_0) - a_1'(\hat{\phi})](\hat{\phi} - \phi_0),$$

where $\hat{\phi}$ is the MLE of ϕ.

The improved gradient statistic to test the null hypothesis $\mathcal{H}_0 : \phi = \phi_0$ reduces to

$$S_T^* = S_T\left[1 - \left(c + bS_T + aS_T^2\right)\right],$$

where $a = A_3/180$, $b = (A_2 - 2A_3)/36$, $c = (A_1 - A_2 + A_3)/12$, and

$$A_1 = -\frac{3p(p+2)}{nd_{(2)}} - \frac{3(3pd_{(3)} - 4d_{(4)})}{nd_{(2)}^2} - \frac{18d_{(3)}^2}{nd_{(2)}^3},$$

$$A_2 = -\frac{3(pd_{(3)} - d_{(4)})}{nd_{(2)}^2} - \frac{33d_{(3)}^2}{4nd_{(2)}^3}, \quad A_3 = -\frac{5d_{(3)}^2}{4nd_{(2)}^3}.$$

Notice that the formulas for the As are very simple, depend on $\boldsymbol{X}$ only through its rank and do not depend on the unknown parameter vector $\boldsymbol{\beta}$. The As are all evaluated at ϕ_0. Under the null hypothesis, the adjusted statistic S_{T}^* has a χ_1^2 distribution up to an error of order $O(n^{-3/2})$.

3.6.4 Real Data Illustration

We consider the usual gradient statistic and the improved gradient statistic for testing hypotheses in the class of GLMs in a real data set. The data correspond to an experiment to study the size of squid eaten by sharks and tuna. The study involved measurements taken on $n = 22$ squids, and the data are reported in Freund [26]. The variables considered in the study are: weight in pounds, rostral length (x_2), wing length (x_3), rostral to notch length (x_4), notch to wing length (x_5), and width (x_6) in inches. Notice that the regressor variables are characteristics of the beak, mouth or wing of the squids.

We consider the systematic component

$$\log \mu_l = \beta_1 x_{1l} + \beta_2 x_{2l} + \beta_3 x_{3l} + \beta_4 x_{4l} + \beta_5 x_{5l} + \beta_6 x_{6l}, \tag{3.13}$$

where $l = 1, \ldots, 22$, $x_{1l} = 1$, and $\phi > 0$ is assumed unknown and it is the same for all observations. We assume a gamma distribution for the response variable Y (weight), that is, $Y_l \sim \text{Gamma}(\mu_l, \phi)$, $l = 1, \ldots, 22$. The MLEs of the regression parameters and the asymptotic standard errors (SE) are listed in Table 3.7. We constructed the normal probability plot with generated envelopes for the deviance component residuals (see, eg, [17]) of the regression model (3.13) fitted to the data (not shown). It reveals that the assumption of the gamma distribution for the data seems suitable, since there are no observations falling outside the envelope.

Table 3.7 MLEs and Asymptotic Standard Errors

Parameter	Estimate	SE
β_1	−2.2899	0.2001
β_2	0.4027	0.5515
β_3	−0.4362	0.5944
β_4	1.2916	1.3603
β_5	1.9420	0.7844
β_6	2.1394	1.0407
ϕ	44.001	13.217

Suppose the interest lies in testing the null hypothesis

$$\mathcal{H}_0 : \beta_4 = \beta_5 = 0,$$

against a two-sided alternative hypothesis; that is, we want to verify whether there is a significant joint effect of rostral to notch length and notch to wing length on the mean weight of squids. For testing $\mathcal{H}_0$, the observed values of S_{T} and S_{T}^* (the corresponding p-values between parentheses) are 5.1193 (0.0773) and 4.3239 (0.1151), respectively. It is noteworthy that one rejects the null hypothesis at the 10% nominal level when the inference is based on the usual gradient test. However, a different decision is reached when the improved gradient test is employed. Recall from the simulation study that the unmodified test is size distorted when the sample is small (here, $n = 22$), and it is considerably affected by the number of tested parameters in the null hypothesis (here, $q = 2$) and by the number of regression parameters in the model (here, $p = 6$), which leads us to mistrust the inference delivered by this test. Therefore, on the basis of the adjusted gradient test, the null hypothesis $\mathcal{H}_0 : \beta_4 = \beta_5 = 0$ should not be rejected at the 10% nominal level.

Let $\mathcal{H}_0 : \beta_2 = \beta_3 = \beta_4 = 0$ be now the null hypothesis of interest, which means the exclusion of the covariates rostral length, wing length and rostral to notch length from the regression model (3.13). The null hypothesis is not rejected at the 10% nominal level by both tests, but we note that the Bartlett-corrected gradient test yields larger p-value. The test statistics values are $S_{\mathrm{T}} = 1.2876$ and $S_{\mathrm{T}}^* = 1.0044$, the corresponding p-values being 0.7321 and 0.8002. Both tests reject the null hypothesis $\mathcal{H}_0 : \beta_2 = \beta_3 = \beta_4 = \beta_5 = 0$ (p-values smaller than 0.01).

We proceed by removing x_2, x_3, and x_4 from model (3.13). We then estimate

$$\log \mu_l = \beta_1 x_{1l} + \beta_5 x_{5l} + \beta_6 x_{6l}, \quad l = 1, \ldots, 22. \tag{3.14}$$

The parameter estimates are (asymptotic standard errors in parentheses): $\hat{\beta}_1 = -2.1339\,(0.1358)$, $\hat{\beta}_5 = 2.1428\,(0.3865)$, $\hat{\beta}_6 = 2.9749\,(0.5888)$, and $\hat{\phi} = 41.4440\,(12.446)$. The null hypotheses $\mathcal{H}_0 : \beta_j = 0$ $(j = 5, 6)$ and

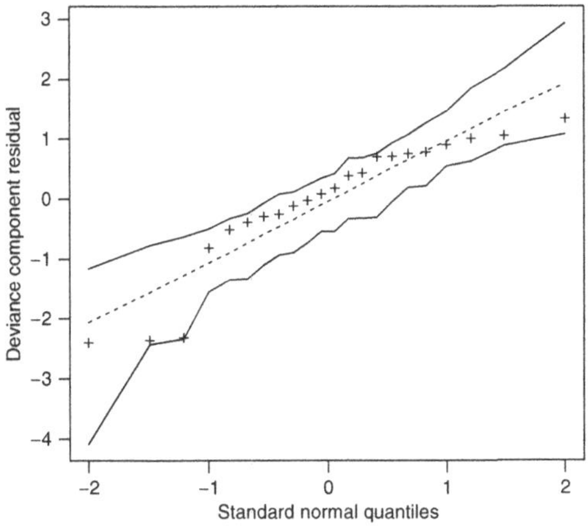

Fig. 3.4 Normal probability plot with envelope for the final model (3.14).

$\mathcal{H}_0 : \beta_5 = \beta_6 = 0$ are rejected by the two tests (unmodified and modified) at any usual significance levels. Hence, the estimated model is

$$\hat{\mu}_l = \mathrm{e}^{-2.1339+2.1428x_{5l}+2.9749x_{6l}}, \quad l = 1, \ldots, 22.$$

Fig. 3.4 displays the normal probability plot with generated envelopes for the deviance component residuals of the final model (3.14) fitted to the data. Notice that there are no observations falling outside the envelope, and hence the final model seems suitable for fitting the data at hand.

CHAPTER 4

The Gradient Statistic Under Model Misspecification

4.1 INTRODUCTION

Let $x_1, \ldots, x_n$ be a set of n independent and identically distributed observations of a random variable X from some unknown probability density function $g(x) = g(x; \boldsymbol{\theta})$, which depends on a parameter vector $\boldsymbol{\theta} = (\theta_1, \ldots, \theta_p)^\top$ of dimension p. We assume that $\boldsymbol{\theta} \in \boldsymbol{\Theta}$, where $\boldsymbol{\Theta} \subseteq \mathbb{R}^p$ is an open subset of the Euclidean space. On the basis of the probability density function $g(x)$, let

$$\ell_n(\boldsymbol{\theta}) = \sum_{l=1}^{n} \log g(x_l; \boldsymbol{\theta})$$

denote the total log-likelihood function,

$$\boldsymbol{U}_n(\boldsymbol{\theta}) = \frac{\partial \ell_n(\boldsymbol{\theta})}{\partial \boldsymbol{\theta}}$$

be the total score vector, and

$$\boldsymbol{K}_n(\boldsymbol{\theta}) = \mathbb{E}[\boldsymbol{U}_n(\boldsymbol{\theta})\boldsymbol{U}_n(\boldsymbol{\theta})^\top] = -\mathbb{E}\left[\frac{\partial \boldsymbol{U}_n(\boldsymbol{\theta})}{\partial \boldsymbol{\theta}^\top}\right] = n\boldsymbol{K}(\boldsymbol{\theta})$$

be the total Fisher information matrix, where $\boldsymbol{K}(\boldsymbol{\theta})$ is the per observation Fisher information matrix. Let $\boldsymbol{K}_n(\boldsymbol{\theta})^{-1} = n^{-1}\boldsymbol{K}(\boldsymbol{\theta})^{-1}$ be the inverse of $\boldsymbol{K}_n(\boldsymbol{\theta})$. Suppose the interest lies in testing the null hypothesis $\mathcal{H}_0 : \boldsymbol{\theta} = \boldsymbol{\theta}_0$ against the alternative hypothesis $\mathcal{H}_a : \boldsymbol{\theta} \neq \boldsymbol{\theta}_0$, where $\boldsymbol{\theta}_0$ is a specified vector. The likelihood ratio (S_{LR}), Wald (S_W), score (S_R), and gradient (S_T) statistics for testing $\mathcal{H}_0$ are given by

$$S_{LR} = 2\left[\ell_n(\hat{\boldsymbol{\theta}}) - \ell_n(\boldsymbol{\theta}_0)\right],$$

The Gradient Test. http://dx.doi.org/10.1016/B978-0-12-803596-2.00004-1

$$S_{\mathrm{W}} = n(\hat{\boldsymbol{\theta}} - \boldsymbol{\theta}_0)^\top \boldsymbol{K}(\hat{\boldsymbol{\theta}})(\hat{\boldsymbol{\theta}} - \boldsymbol{\theta}_0),$$
$$S_{\mathrm{R}} = n^{-1}\boldsymbol{U}_n(\boldsymbol{\theta}_0)^\top \boldsymbol{K}(\boldsymbol{\theta}_0)^{-1}\boldsymbol{U}_n(\boldsymbol{\theta}_0),$$
$$S_{\mathrm{T}} = \boldsymbol{U}_n(\boldsymbol{\theta}_0)^\top(\hat{\boldsymbol{\theta}} - \boldsymbol{\theta}_0),$$

where $\hat{\boldsymbol{\theta}} = (\hat{\theta}_1, \ldots, \hat{\theta}_p)^\top$ is the MLE of $\boldsymbol{\theta} = (\theta_1, \ldots, \theta_p)^\top$, which is obtained from $\boldsymbol{U}_n(\hat{\boldsymbol{\theta}}) = \mathbf{0}_p$. As earlier noted, the gradient statistic S_{T} has a very simple form and does not involve the information matrix, neither expected nor observed, unlike S_{W} and S_{R}. It is simply the inner product of the score vector evaluated at the null hypothesis $\mathcal{H}_0$ and the difference between the unrestricted and restricted MLEs of $\boldsymbol{\theta}$.

The gradient statistic has been considered in several special classes of parametric models for testing hypotheses on the model parameters, however, only under model correctness. Readers are referred to the works of Lemonte [16, 27, 28], Lemonte and Ferrari [8, 29–31], Vargas et al. [25], Medeiros et al. [32], Ferrari and Pinheiro [33], among others. In this chapter, which is based on Lemonte [34], we present a robust gradient statistic that can be used for testing hypotheses under model misspecification. Some examples as well as Monte Carlo simulations are also presented and discussed. The robust gradient statistic can be easily computed and can be an interesting alternative to the robust versions of the Wald and score statistics. Finally, it should be mentioned that the concern of this chapter is solely with asymptotic results and hence no small-sample results are included. For some results on higher-order asymptotics under model misspecification, the reader is referred to the works of Viraswami and Reid [35, 36], Stafford [37], and Royall and Tsou [38].

4.2 THE ROBUST GRADIENT STATISTIC

In this section, we assume that we have observed a set of n observations $x_1, \ldots, x_n$ from some unknown probability density function $g(x)$, but the data will be summarized using a parametric model $\{f(x;\boldsymbol{\theta}); \boldsymbol{\theta} \in \boldsymbol{\Theta}\}$. If the log-likelihood function for $\boldsymbol{\theta}$ based on the sample $x_1, \ldots, x_n$ is now given by

$$\ell_n(\boldsymbol{\theta}) = \sum_{l=1}^{n} \log f(x_l;\boldsymbol{\theta}), \tag{4.1}$$

then $\boldsymbol{\theta}(g)$ is defined in Kent [39] to be the value $\boldsymbol{\theta} \in \boldsymbol{\Theta}$ which maximizes the Fraser's [40] information of $f(x;\boldsymbol{\theta})$ under $g(x)$: $F_I(\boldsymbol{\theta}) = \int \log f(x;\boldsymbol{\theta}) g(x)\, dx$;

that is, $F_I(\boldsymbol{\theta}(g)) = \max\{F_I(\boldsymbol{\theta}); \boldsymbol{\theta} \in \boldsymbol{\Theta}\}$. Thus, $\boldsymbol{\theta}(g)$ is the theoretical analogue of the MLE of $\boldsymbol{\theta}$. Equivalently, $\boldsymbol{\theta}(g)$ is the limit in probability of the MLE $\hat{\boldsymbol{\theta}}$ of $\boldsymbol{\theta}$, estimated under the model $\{f(x; \boldsymbol{\theta}); \boldsymbol{\theta} \in \boldsymbol{\Theta}\}$ when the probability density function is in fact $g(x)$.

In what follows, we are interested in testing the null hypothesis $\mathcal{H}_0 : \boldsymbol{\theta}(g) = \boldsymbol{\theta}_0$ against the alternative hypothesis $\mathcal{H}_a : \boldsymbol{\theta}(g) \neq \boldsymbol{\theta}_0$. If the density $g(x)$ is a member of the parametric family $\{f(x; \boldsymbol{\theta}); \boldsymbol{\theta} \in \boldsymbol{\Theta}\}$, then the test statistics S_{LR}, S_{W}, S_{R}, and S_{T} presented earlier might be used. In the type of model misspecification we are considering, the true underlying density $g(x)$ does not belong to the parametric family $\{f(x; \boldsymbol{\theta}); \boldsymbol{\theta} \in \boldsymbol{\Theta}\}$, but still satisfies $\boldsymbol{\theta}(g) = \boldsymbol{\theta}_0$. We follow Kent [39] and assume that the parametric family $\{f(x; \boldsymbol{\theta}); \boldsymbol{\theta} \in \boldsymbol{\Theta}\}$ is not regarded as a true model for the data, however, the parameter $\boldsymbol{\theta}$ is regarded as a convenient way to summarize the data and hence has a meaningful interpretation outside the parametric family. The hypothesis $\boldsymbol{\theta}(g) = \boldsymbol{\theta}_0$ is of interest whenever $\boldsymbol{\theta}(g)$ is directly interpretable under $g(x)$, for example, as an expected value.

Kent [39] showed that under the type of model misspecification above and under the assumption that $\hat{\boldsymbol{\theta}}$ is a consistent estimate of $\boldsymbol{\theta}_0$, the first-order equivalence of the statistics S_{LR}, S_{W}, and S_{R} still holds, but their distributions depart from the familiar likelihood result and involve a linear combination of independent χ^2 variates with coefficients given by the eigenvalues of the matrix $\boldsymbol{H}(\boldsymbol{\theta}_0)^{-1}\boldsymbol{J}(\boldsymbol{\theta}_0)$, where

$$\boldsymbol{J}(\boldsymbol{\theta}) = \int_{-\infty}^{\infty} \boldsymbol{u}(\boldsymbol{\theta})\boldsymbol{u}(\boldsymbol{\theta})^\top g(x)\, dx, \quad \boldsymbol{H}(\boldsymbol{\theta}) = -\int_{-\infty}^{\infty} \frac{\partial \boldsymbol{u}(\boldsymbol{\theta})}{\partial \boldsymbol{\theta}^\top} g(x)\, dx,$$

and

$$\boldsymbol{u}(\boldsymbol{\theta}) = \frac{\partial}{\partial \boldsymbol{\theta}} \log f(x; \boldsymbol{\theta}) = \left(\frac{\partial}{\partial \theta_j} \log f(x; \boldsymbol{\theta}) \right)_{j=1,\ldots,p}.$$

The matrix $\boldsymbol{J}(\boldsymbol{\theta})$ is the covariance of the score function and the matrix $\boldsymbol{H}(\boldsymbol{\theta})$ is the expected value of minus the score function derivative. Note that $\boldsymbol{J}(\boldsymbol{\theta}) \neq \boldsymbol{H}(\boldsymbol{\theta})$, that is, the first Bartlett identity does not hold.

Kent [39] suggested alternative test statistics which are robust in the sense that, under the null hypothesis, they have an asymptotic χ^2 distribution with p degrees of freedom. A robust version of the Wald test statistic is

$$W = n(\hat{\boldsymbol{\theta}} - \boldsymbol{\theta}_0)^\top \boldsymbol{H}(\boldsymbol{\theta}_0)\boldsymbol{J}(\boldsymbol{\theta}_0)^{-1}\boldsymbol{H}(\boldsymbol{\theta}_0)(\hat{\boldsymbol{\theta}} - \boldsymbol{\theta}_0),$$

whereas a robust version of the Rao score test statistic takes the form

$$R = n^{-1}\boldsymbol{u}_n(\boldsymbol{\theta}_0)^\top \boldsymbol{J}(\boldsymbol{\theta}_0)^{-1}\boldsymbol{u}_n(\boldsymbol{\theta}_0),$$

where $\boldsymbol{u}_n(\boldsymbol{\theta})$ is obtained from the log-likelihood function (4.1) in the form

$$\boldsymbol{u}_n(\boldsymbol{\theta}) = \frac{\partial \ell_n(\boldsymbol{\theta})}{\partial \boldsymbol{\theta}} = \sum_{l=1}^{n} \frac{\partial}{\partial \boldsymbol{\theta}} \log f(x_l; \boldsymbol{\theta}).$$

There is no robust version of the likelihood ratio statistic itself. From Kent [39], the robust statistics W and R have an asymptotic χ_p^2 distribution under the null hypothesis and under the model misspecification that we are considering.

We now present a robust version of the gradient statistic under model misspecification. First, let $\boldsymbol{L}(\boldsymbol{\theta})$ be a $p \times p$ matrix such that $\boldsymbol{L}(\boldsymbol{\theta})^\top \boldsymbol{L}(\boldsymbol{\theta}) = \boldsymbol{J}(\boldsymbol{\theta})$. We can express W and R as

$$\begin{aligned} W &= n(\hat{\boldsymbol{\theta}} - \boldsymbol{\theta}_0)^\top \boldsymbol{H}(\boldsymbol{\theta}_0)(\boldsymbol{L}(\boldsymbol{\theta}_0)^\top \boldsymbol{L}(\boldsymbol{\theta}_0))^{-1}\boldsymbol{H}(\boldsymbol{\theta}_0)(\hat{\boldsymbol{\theta}} - \boldsymbol{\theta}_0) \\ &= [n^{1/2}(\boldsymbol{L}(\boldsymbol{\theta}_0)^{-1})^\top \boldsymbol{H}(\boldsymbol{\theta}_0)(\hat{\boldsymbol{\theta}} - \boldsymbol{\theta}_0)]^\top \\ &\quad \times [n^{1/2}(\boldsymbol{L}(\boldsymbol{\theta}_0)^\top)^{-1}\boldsymbol{H}(\boldsymbol{\theta}_0)(\hat{\boldsymbol{\theta}} - \boldsymbol{\theta}_0)], \end{aligned}$$

$$\begin{aligned} R &= n^{-1}\boldsymbol{u}_n(\boldsymbol{\theta}_0)^\top (\boldsymbol{L}(\boldsymbol{\theta}_0)^\top \boldsymbol{L}(\boldsymbol{\theta}_0))^{-1}\boldsymbol{u}_n(\boldsymbol{\theta}_0) \\ &= [n^{-1/2}(\boldsymbol{L}(\boldsymbol{\theta}_0)^{-1})^\top \boldsymbol{u}_n(\boldsymbol{\theta}_0)]^\top [n^{-1/2}(\boldsymbol{L}(\boldsymbol{\theta}_0)^\top)^{-1}\boldsymbol{u}_n(\boldsymbol{\theta}_0)]. \end{aligned}$$

We define

$$\boldsymbol{P}_1 = n^{-1/2}(\boldsymbol{L}(\boldsymbol{\theta}_0)^{-1})^\top \boldsymbol{u}_n(\boldsymbol{\theta}_0),$$

and

$$\boldsymbol{P}_2 = n^{1/2}(\boldsymbol{L}(\boldsymbol{\theta}_0)^\top)^{-1}\boldsymbol{H}(\boldsymbol{\theta}_0)(\hat{\boldsymbol{\theta}} - \boldsymbol{\theta}_0).$$

From Kent [39], it follows that

$$\begin{aligned} n^{-1/2}\boldsymbol{u}_n(\boldsymbol{\theta}_0) &\overset{a}{\sim} \mathcal{N}_p(\boldsymbol{0}_p, \boldsymbol{J}(\boldsymbol{\theta}_0)), \\ n^{1/2}(\hat{\boldsymbol{\theta}} - \boldsymbol{\theta}_0) &\overset{a}{\sim} \mathcal{N}_p(\boldsymbol{0}_p, \boldsymbol{H}(\boldsymbol{\theta}_0)^{-1}\boldsymbol{J}(\boldsymbol{\theta}_0)\boldsymbol{H}(\boldsymbol{\theta}_0)^{-1}), \end{aligned}$$

when n is large. So, using these results we have that $\boldsymbol{P}_1 \overset{a}{\sim} \mathcal{N}_p(\boldsymbol{0}_p, \boldsymbol{I}_p)$ and $\boldsymbol{P}_2 \overset{a}{\sim} \mathcal{N}_p(\boldsymbol{0}_p, \boldsymbol{I}_p)$. Notice that the inner product of these standardized vectors has asymptotically χ^2 distribution with p degrees of freedom:

$$\begin{aligned} \boldsymbol{P}_1^\top \boldsymbol{P}_2 &= [n^{-1/2}(\boldsymbol{L}(\boldsymbol{\theta}_0)^{-1})^\top \boldsymbol{u}_n(\boldsymbol{\theta}_0)]^\top [n^{1/2}(\boldsymbol{L}(\boldsymbol{\theta}_0)^\top)^{-1}\boldsymbol{H}(\boldsymbol{\theta}_0)(\hat{\boldsymbol{\theta}} - \boldsymbol{\theta}_0)] \\ &= \boldsymbol{u}_n(\boldsymbol{\theta}_0)^\top \boldsymbol{J}(\boldsymbol{\theta}_0)^{-1}\boldsymbol{H}(\boldsymbol{\theta}_0)(\hat{\boldsymbol{\theta}} - \boldsymbol{\theta}_0). \end{aligned}$$

Note that the above procedure is analogous to that of Terrell [6], who proposed the gradient statistic under model correctness.

We have the following definition.

Definition 4.1. The gradient statistic for testing the null hypothesis $\mathcal{H}_0$: $\boldsymbol{\theta}(g) = \boldsymbol{\theta}_0$ under model misspecification is

$$T = \boldsymbol{u}_n(\boldsymbol{\theta}_0)^\top \boldsymbol{J}(\boldsymbol{\theta}_0)^{-1} \boldsymbol{H}(\boldsymbol{\theta}_0)(\hat{\boldsymbol{\theta}} - \boldsymbol{\theta}_0). \tag{4.2}$$

We initially consider the following theorem.

Theorem 4.1. *The gradient statistic T in Eq. (4.2) has a* χ_p^2 *distribution under the null hypothesis.*

Proof. By making use of the regularity conditions in Kent [38, Section 8], after some algebra, it follows that

$$n^{1/2}(\hat{\boldsymbol{\theta}} - \boldsymbol{\theta}_0) = n^{-1/2}\boldsymbol{H}(\boldsymbol{\theta}_0)^{-1}\boldsymbol{u}_n(\boldsymbol{\theta}_0) + O_p(n^{-1/2}).$$

Now, from the definition of the gradient statistic in Eq. (4.2) and using the fact that $n^{-1/2}\boldsymbol{u}_n(\boldsymbol{\theta}_0) = O_p(1)$, we have

$$\begin{aligned} T &= \boldsymbol{u}_n(\boldsymbol{\theta}_0)^\top \boldsymbol{J}(\boldsymbol{\theta}_0)^{-1} \boldsymbol{H}(\boldsymbol{\theta}_0)(\hat{\boldsymbol{\theta}} - \boldsymbol{\theta}_0) \\ &= n^{-1/2}\boldsymbol{u}_n(\boldsymbol{\theta}_0)^\top \boldsymbol{J}(\boldsymbol{\theta}_0)^{-1} \boldsymbol{H}(\boldsymbol{\theta}_0)[n^{1/2}(\hat{\boldsymbol{\theta}} - \boldsymbol{\theta}_0)] \\ &= n^{-1/2}\boldsymbol{u}_n(\boldsymbol{\theta}_0)^\top \boldsymbol{J}(\boldsymbol{\theta}_0)^{-1} \boldsymbol{H}(\boldsymbol{\theta}_0) \\ &\quad \times [n^{-1/2}\boldsymbol{H}(\boldsymbol{\theta}_0)^{-1}\boldsymbol{u}_n(\boldsymbol{\theta}_0) + O_p(n^{-1/2})] \\ &= n^{-1}\boldsymbol{u}_n(\boldsymbol{\theta}_0)^\top \boldsymbol{J}(\boldsymbol{\theta}_0)^{-1}\boldsymbol{u}_n(\boldsymbol{\theta}_0) + O_p(n^{-1/2}); \end{aligned}$$

that is, $T = R + O_p(n^{-1/2}) = R + o_p(1)$. Therefore, once R has an asymptotic χ_p^2 distribution under the null hypothesis and under model misspecification, the result follows. □

The following theorem is related to a peculiarity of the statistic defined in Eq. (4.2); that is, this statistic is not transparently non-negative, even though it must be so asymptotically.

Theorem 4.2. *If the log-likelihood function* (4.1) *is concave and differentiable at some* $\boldsymbol{\theta}_0 \in \boldsymbol{\Theta}$, *then* $T \geq 0$.

Proof. From Kent [39], we have that $\boldsymbol{J}(\boldsymbol{\theta})$ and $\boldsymbol{H}(\boldsymbol{\theta})$ are positive definite matrices and hence $\boldsymbol{J}(\boldsymbol{\theta})^{-1}\boldsymbol{H}(\boldsymbol{\theta})$ is a positive definite matrix. Therefore, on the basis of Theorem 2 in Terrell [6], the result holds. □

In general, $\boldsymbol{J}(\boldsymbol{\theta})$ and $\boldsymbol{H}(\boldsymbol{\theta})$ will not be known, but we can replace these matrices by any consistent estimates in practice. As suggested by Kent [39], we can consider the consistent estimates

$$\hat{\boldsymbol{J}}_n(\boldsymbol{\theta}) = \frac{1}{n}\sum_{l=1}^{n} \boldsymbol{u}^{(l)}(\boldsymbol{\theta})\boldsymbol{u}^{(l)}(\boldsymbol{\theta})^\top, \quad \hat{\boldsymbol{H}}_n(\boldsymbol{\theta}) = -\frac{1}{n}\sum_{l=1}^{n} \frac{\partial \boldsymbol{u}^{(l)}(\boldsymbol{\theta})}{\partial \boldsymbol{\theta}^\top},$$

where

$$\boldsymbol{u}^{(l)}(\boldsymbol{\theta}) = \frac{\partial}{\partial \boldsymbol{\theta}} \log f(x_l; \boldsymbol{\theta}) = \left(\frac{\partial}{\partial \theta_j} \log f(x_l; \boldsymbol{\theta})\right)_{j=1,\ldots,p}.$$

Notice that $\boldsymbol{u}_n(\boldsymbol{\theta}) = \sum_{l=1}^{n} \boldsymbol{u}^{(l)}(\boldsymbol{\theta})$. The robust statistic becomes

$$T^* = \boldsymbol{u}_n(\boldsymbol{\theta}_0)^\top \hat{\boldsymbol{J}}_n(\boldsymbol{\theta}_0)^{-1}\hat{\boldsymbol{H}}_n(\boldsymbol{\theta}_0)(\hat{\boldsymbol{\theta}} - \boldsymbol{\theta}_0).$$

From Kent [39], we have that

$$\hat{\boldsymbol{H}}_n(\boldsymbol{\theta}) = \boldsymbol{H}(\boldsymbol{\theta}) + O_p(n^{-1/2}), \quad \hat{\boldsymbol{J}}_n(\boldsymbol{\theta}) = \boldsymbol{J}(\boldsymbol{\theta}) + O_p(n^{-1/2}).$$

So, the robust statistic T^* can be expressed as

$$T^* = T + O_p(n^{-1/2}) = T + o_p(1).$$

Therefore, under the null hypothesis, the limiting distribution of T^* is also χ^2 with p degrees of freedom.

Note that the quantity defined by $\widehat{\mathbb{COV}}(\hat{\boldsymbol{\theta}}) = \hat{\boldsymbol{H}}_n(\hat{\boldsymbol{\theta}})^{-1}\hat{\boldsymbol{J}}_n(\hat{\boldsymbol{\theta}})\hat{\boldsymbol{H}}_n(\hat{\boldsymbol{\theta}})^{-1}$ is known as "sandwich" estimator, and it is a consistent estimator of the asymptotic variance-covariance matrix of $\hat{\boldsymbol{\theta}}$. It occurs naturally in the theory of M-estimation. Also, the matrix $\hat{\boldsymbol{H}}_n(\hat{\boldsymbol{\theta}})$ is often called the observed Fisher information matrix. It should be mentioned that Heritier and Ronchetti [41] derived versions of the Wald and Rao score statistics that are robust in the sense of having bounded influence functions. Their definition of robustness is different from that used here and in Kent [39]. The robustness studied here corresponds to what Heritier and Ronchetti [41] call *robustness of validity* (ie, the level of a test should be stable under small arbitrary departures from the null hypothesis) and *robustness of efficiency* (ie, the test should still have a good power under small arbitrary departures from specified alternatives).

Now, consider the partition $\boldsymbol{\theta} = (\boldsymbol{\psi}^\top, \boldsymbol{\lambda}^\top)^\top$ and similarly $\boldsymbol{\theta}(g) = (\boldsymbol{\psi}(g)^\top, \boldsymbol{\lambda}(g)^\top)^\top$. Let $\boldsymbol{\psi}$ and $\boldsymbol{\lambda}$ be q- and $(p-q)$-dimensional, respectively.

We wish to test the composite null hypothesis $\mathcal{H}_0 : \boldsymbol{\psi}(g) = \boldsymbol{\psi}_0$ against the alternative hypothesis $\mathcal{H}_a : \boldsymbol{\psi}(g) \neq \boldsymbol{\psi}_0$, where $\boldsymbol{\psi}_0$ is a specified value of $\boldsymbol{\psi}$, and $\boldsymbol{\lambda}$ acts a nuisance parameter vector. The partition of $\boldsymbol{\theta} = (\boldsymbol{\psi}^\top, \boldsymbol{\lambda}^\top)^\top$ induces the corresponding partitions:

$$\boldsymbol{H}(\boldsymbol{\theta}) = \begin{bmatrix} \boldsymbol{H}_{\psi\psi}(\boldsymbol{\theta}) & \boldsymbol{H}_{\psi\lambda}(\boldsymbol{\theta}) \\ \boldsymbol{H}_{\lambda\psi}(\boldsymbol{\theta}) & \boldsymbol{H}_{\lambda\lambda}(\boldsymbol{\theta}) \end{bmatrix}, \quad \boldsymbol{J}(\boldsymbol{\theta}) = \begin{bmatrix} \boldsymbol{J}_{\psi\psi}(\boldsymbol{\theta}) & \boldsymbol{J}_{\psi\lambda}(\boldsymbol{\theta}) \\ \boldsymbol{J}_{\lambda\psi}(\boldsymbol{\theta}) & \boldsymbol{J}_{\lambda\lambda}(\boldsymbol{\theta}) \end{bmatrix},$$

$$\boldsymbol{H}(\boldsymbol{\theta})^{-1} = \begin{bmatrix} \boldsymbol{H}^{\psi\psi}(\boldsymbol{\theta}) & \boldsymbol{H}^{\psi\lambda}(\boldsymbol{\theta}) \\ \boldsymbol{H}^{\lambda\psi}(\boldsymbol{\theta}) & \boldsymbol{H}^{\lambda\lambda}(\boldsymbol{\theta}) \end{bmatrix}, \quad \boldsymbol{J}(\boldsymbol{\theta})^{-1} = \begin{bmatrix} \boldsymbol{J}^{\psi\psi}(\boldsymbol{\theta}) & \boldsymbol{J}^{\psi\lambda}(\boldsymbol{\theta}) \\ \boldsymbol{J}^{\lambda\psi}(\boldsymbol{\theta}) & \boldsymbol{J}^{\lambda\lambda}(\boldsymbol{\theta}) \end{bmatrix}.$$

Similarly, we can partition

$$\boldsymbol{u}_n(\boldsymbol{\theta}) = \begin{pmatrix} \boldsymbol{u}_{n\psi}(\boldsymbol{\theta}) \\ \boldsymbol{u}_{n\lambda}(\boldsymbol{\theta}) \end{pmatrix} = \begin{pmatrix} \sum_{l=1}^{n} \frac{\partial}{\partial \boldsymbol{\psi}} \log f(x_l; \boldsymbol{\theta}) \\ \sum_{l=1}^{n} \frac{\partial}{\partial \boldsymbol{\lambda}} \log f(x_l; \boldsymbol{\theta}) \end{pmatrix}.$$

Let $\hat{\boldsymbol{\theta}} = (\hat{\boldsymbol{\psi}}^\top, \hat{\boldsymbol{\lambda}}^\top)^\top$ and $\tilde{\boldsymbol{\theta}} = (\boldsymbol{\psi}_0^\top, \tilde{\boldsymbol{\lambda}}^\top)^\top$ denote the unrestricted and restricted MLEs of $\boldsymbol{\theta} = (\boldsymbol{\psi}^\top, \boldsymbol{\lambda}^\top)^\top$, respectively. These estimates satisfy $\boldsymbol{u}_n(\hat{\boldsymbol{\theta}}) = \boldsymbol{0}_p$ and $\boldsymbol{u}_{n\lambda}(\tilde{\boldsymbol{\theta}}) = \boldsymbol{0}_{p-q}$. The robust gradient statistic takes the form

$$T = \boldsymbol{u}_{n\psi}(\tilde{\boldsymbol{\theta}})^\top [\boldsymbol{J}(\tilde{\boldsymbol{\theta}})^{-1} \boldsymbol{H}(\tilde{\boldsymbol{\theta}})]_{\psi\psi} (\hat{\boldsymbol{\psi}} - \boldsymbol{\psi}_0),$$

where $[A]_{\psi\psi}$ denotes a $q \times q$ matrix related to the parameter vector of interest $\boldsymbol{\psi}$ obtained from $\boldsymbol{A}$. We can also express T in the form

$$T = \boldsymbol{u}_{n\psi}(\tilde{\boldsymbol{\theta}})^\top [\boldsymbol{H}(\tilde{\boldsymbol{\theta}})^{-1} \boldsymbol{J}(\tilde{\boldsymbol{\theta}})]_{\psi\psi}^{-1} (\hat{\boldsymbol{\psi}} - \boldsymbol{\psi}_0).$$

If $\boldsymbol{H}(\boldsymbol{\theta})$ is block diagonal, that is, $\boldsymbol{H}_{\psi\lambda}(\boldsymbol{\theta}) = \boldsymbol{0}_{q,p-q}$, then

$$T = \boldsymbol{u}_{n\psi}(\tilde{\boldsymbol{\theta}})^\top \boldsymbol{J}_{\psi\psi}(\tilde{\boldsymbol{\theta}})^{-1} \boldsymbol{H}_{\psi\psi}(\tilde{\boldsymbol{\theta}}) (\hat{\boldsymbol{\psi}} - \boldsymbol{\psi}_0).$$

Under the null hypothesis, the limiting distribution of the statistic T is χ^2 with q degrees of freedom. The asymptotic distribution of T can be obtained by noting that

$$n^{1/2}(\hat{\boldsymbol{\psi}} - \boldsymbol{\psi}_0) = n^{-1/2} \boldsymbol{H}_{\psi\psi.\lambda}(\tilde{\boldsymbol{\theta}}) \boldsymbol{u}_{n\psi}(\tilde{\boldsymbol{\theta}}) + O_p(n^{-1/2}),$$

where $\boldsymbol{H}_{\psi\psi.\lambda}(\boldsymbol{\theta}) = (\boldsymbol{H}_{\psi\psi}(\boldsymbol{\theta}) - \boldsymbol{H}_{\psi\lambda}(\boldsymbol{\theta}) \boldsymbol{H}_{\lambda\lambda}(\boldsymbol{\theta})^{-1} \boldsymbol{H}_{\lambda\psi}(\boldsymbol{\theta}))^{-1}$. As remarked before, the matrices $\boldsymbol{H}(\boldsymbol{\theta})$ and $\boldsymbol{J}(\boldsymbol{\theta})$ can be replaced by the empirical estimates $\hat{\boldsymbol{H}}_n(\boldsymbol{\theta})$ and $\hat{\boldsymbol{J}}_n(\boldsymbol{\theta})$, respectively.

The empirical estimates of $\boldsymbol{H}(\boldsymbol{\theta})$ and $\boldsymbol{J}(\boldsymbol{\theta})$ given by $\hat{\boldsymbol{H}}_n(\boldsymbol{\theta})$ and $\hat{\boldsymbol{J}}_n(\boldsymbol{\theta})$, respectively, may be imprecise when the sample size n is not sufficiently large compared to the dimension of the parameter vector $\boldsymbol{\theta}$. As a consequence of a poor estimation of $\boldsymbol{H}(\boldsymbol{\theta})$ and $\boldsymbol{J}(\boldsymbol{\theta})$, the robust statistic T^* may be numerically unstable. Varin et al. [42] provide recent discussions of computational aspects related to the empirical estimates $\hat{\boldsymbol{H}}_n(\boldsymbol{\theta})$ and $\hat{\boldsymbol{J}}_n(\boldsymbol{\theta})$ in the context of composite marginal likelihoods and indicate alternative estimates obtained, for example, using the jackknife method. Finally, although the notion misspecification is commonly used for the situation where the sample distribution does not belong to the parametric family, it also applies in the case where no simple appropriate parametric family can be found. Then an approximation of the unknown density by a member of the parametric family is estimated. This topic is worth investigating.

4.3 EXAMPLES

We shall now present some examples.

Example 4.1 (One-parameter exponential family). Let $x_1, \ldots, x_n$ be a sample of size n from a continuous distribution. We shall assume that the data come from a continuous distribution which belongs to the one-parameter exponential family of distributions with density

$$f(x;\theta) = \xi(\theta)^{-1} \exp[-\alpha(\theta)d(x) + v(x)],$$

where $\alpha(\cdot)$, $v(\cdot)$, $d(\cdot)$, and $\xi(\cdot)$ are known functions. We wish to test the null hypothesis $\mathcal{H}_0 : \theta = \theta_0$ against $\mathcal{H}_a : \theta \neq \theta_0$, where θ_0 is a fixed value. The statistic T^* can be expressed in the form

$$T^* = n\left(\frac{\theta_0 - \hat{\theta}}{\xi(\theta_0)}\right)\left\{\frac{\left[\xi'(\theta_0) + \bar{d}\,\xi(\theta_0)\alpha'(\theta_0)\right]\left[\xi(\theta_0)\xi''(\theta_0) - \xi'(\theta_0)^2\right]}{\left[\xi'(\theta_0)^2 + 2\,\bar{d}\,\xi(\theta_0)\xi'(\theta_0)\alpha'(\theta_0) + \bar{d}_2\,\xi(\theta_0)^2\alpha'(\theta_0)^2\right]}\right.$$
$$\left. + \frac{\bar{d}\,\xi(\theta_0)^2\alpha''(\theta_0)\left[\xi'(\theta_0) + \bar{d}\,\xi(\theta_0)\alpha'(\theta_0)\right]}{\left[\xi'(\theta_0)^2 + 2\,\bar{d}\,\xi(\theta_0)\xi'(\theta_0)\alpha'(\theta_0) + \bar{d}_2\,\xi(\theta_0)^2\alpha'(\theta_0)^2\right]}\right\},$$

where primes denote derivatives with respect to θ (for instance, $\alpha'(\theta) = d\alpha(\theta)/d\theta$, $\alpha''(\theta) = d^2\alpha(\theta)/d\theta^2$, and so on), $\hat{\theta}$ is the MLE of θ, and

$$\bar{d} = \frac{1}{n}\sum_{l=1}^{n} d(x_l), \quad \bar{d}_2 = \frac{1}{n}\sum_{l=1}^{n} d(x_l)^2.$$

For example, we have $\alpha(\theta) = -\theta/\sigma^2$, $\xi(\theta) = \exp[\theta^2/(2\sigma^2)]$, $d(x) = x$, and $v(x) = -x^2/2 - \log(2\pi\sigma^2)/2$ if we assume that the data come from a normal distribution with mean $\theta \in \mathbb{R}$ and known variance σ^2. So, the statistic T^* becomes

$$T^* = n\,\frac{(\bar{d} - \theta_0)^2}{(\bar{d}_2 - 2\,\bar{d}\,\theta_0 + \theta_0^2)},$$

where $\bar{d} = n^{-1}\sum_{l=1}^{n} x_l$ and $\bar{d}_2 = n^{-1}\sum_{l=1}^{n} x_l^2$. Now, if we assume that the data come from a truncated extreme value distribution with parameter $\theta > 0$, we have $\alpha(\theta) = \theta^{-1}$, $\xi(\theta) = \theta$, $d(x) = \exp(x) - 1$ and $v(x) = x$, and hence the statistic T^* can be reduced to

$$T^* = n\left(\frac{\theta_0 - \bar{d}}{\theta_0}\right)\frac{(\theta_0 - \bar{d})(2\,\bar{d} - \theta_0)}{(\bar{d}_2 - 2\,\bar{d}\,\theta_0 + \theta_0^2)},$$

where $\bar{d} = n^{-1}\sum_{l=1}^{n}[\exp(x_l) - 1]$ and $\bar{d}_2 = n^{-1}\sum_{l=1}^{n}[\exp(x_l) - 1]^2$. On the other hand, assuming that the data come from a gamma model with known shape parameter $\alpha > 0$ and scale parameter $\theta > 0$, we have $\alpha(\theta) = \theta^{-1}$, $\xi(\theta) = \theta^{\alpha}$, $d(x) = x$ and $v(x) = (\alpha - 1)\log x - \log\Gamma(\alpha)$, and therefore

$$T^* = n\left(\frac{\hat{\theta} - \theta_0}{\theta_0}\right)\frac{(\bar{d} - \alpha\,\theta_0)(2\,\bar{d} - \alpha\,\theta_0)}{(\bar{d}_2 - 2\,\bar{d}\,\alpha\,\theta_0 + \alpha^2\theta_0^2)},$$

where $\bar{d} = n^{-1}\sum_{l=1}^{n} x_l$, $\bar{d}_2 = n^{-1}\sum_{l=1}^{n} x_l^2$ and $\hat{\theta} = \alpha^{-1}\bar{d}$. It is evident that several other one-parameter distributions can be considered and expressions for the statistic T^* can be obtained from the general expression above.

Example 4.2 (Birnbaum-Saunders distribution). Suppose that we have a sample $x_1, \ldots, x_n$ of size n from an inverse Gaussian distribution with location parameter $\mu > 0$ and shape parameter $\beta > 0$. We shall assume that the data are from a Birnbaum-Saunders distribution with known scale parameter $\eta > 0$, and shape parameter $\alpha > 0$. We have

$$f(x;\alpha) = \phi\left(\frac{1}{\alpha}\left[\sqrt{\frac{x}{\eta}} - \sqrt{\frac{\eta}{x}}\right]\right)\frac{x^{-3/2}(x+\eta)}{2\alpha\sqrt{\eta}}, \quad x > 0,$$

where $\phi(\cdot)$ is the standard normal probability density function. To test the null hypothesis $\mathcal{H}_0 : \alpha = \alpha_0$, the statistic T^* is given by

$$T^* = n^2\left(\frac{\hat{\alpha} - \alpha_0}{\alpha_0}\right)\frac{(3\,\hat{\alpha}^2/\alpha_0^2 - 1)\,(\hat{\alpha}^2/\alpha_0^2 - 1)}{\sum_{l=1}^{n}\left[\alpha_0^{-2}(x_l\,\eta^{-1} + x_l^{-1}\eta - 2) - 1\right]^2},$$

where $\bar{x} = n^{-1}\sum_{l=1}^{n} x_l$, $\bar{h} = (n^{-1}\sum_{l=1}^{n} x_l^{-1})^{-1}$ and $\hat{\alpha} = (\bar{x}\eta^{-1} + \bar{h}^{-1}\eta - 2)^{1/2}$ is the MLE of α.

Example 4.3 (Log-Birnbaum-Saunders distribution). Let $x_1, \ldots, x_n$ be a sample of size n of the normal distribution with mean $\theta \in \mathbb{R}$ and with known variance, which we assume as unity. We shall assume that the data are from a log-Birnbaum-Saunders distribution (see Section 3.5) with known shape parameter $\phi > 0$ (which for convenience we take as unity), and location parameter $\mu \in \mathbb{R}$. In this case, the log-Birnbaum-Saunders density takes the form

$$f(x;\mu) = (2\pi)^{-1/2}\cosh\left(\frac{x-\mu}{2}\right)\exp\left[-2\sinh^2\left(\frac{x-\mu}{2}\right)\right],$$

where $x \in \mathbb{R}$. We wish to test the null hypothesis $\mathcal{H}_0 : \mu = \mu_0$. We have that

$$u_n(\mu) = \sum_{l=1}^{n}\left\{\sinh(x_l-\mu) - \frac{\sinh[(x_l-\mu)/2]}{2\cosh[(x_l-\mu)/2]}\right\},$$

and hence the MLE $\hat{\mu}$ of μ can be obtained from $u_n(\hat{\mu}) = 0$. Also,

$$\hat{J}_n(\mu) = \frac{1}{n}\sum_{l=1}^{n}\left\{\sinh(x_l-\mu) - \frac{\sinh[(x_l-\mu)/2]}{2\cosh[(x_l-\mu)/2]}\right\}^2,$$

$$\hat{H}_n(\mu) = -\frac{1}{n}\sum_{l=1}^{n}\left\{\frac{1}{4} - \cosh(x_l-\mu) - \frac{\sinh^2[(x_l-\mu)/2]}{4\cosh^2[(x_l-\mu)/2]}\right\}.$$

For testing $\mathcal{H}_0 : \mu = \mu_0$, the robust gradient statistic reduces to

$$T^* = u_n(\mu_0)(\hat{\mu} - \mu_0)\frac{\hat{H}_n(\mu_0)}{\hat{J}_n(\mu_0)}.$$

Example 4.4 (Beta distribution). Let $x_1, \ldots, x_n$ be a sample of size n from a Kumaraswamy distribution with shape parameters $a > 0$ and $b > 0$. We shall assume that the data come from a beta distribution with shape parameters $\alpha > 0$ and $\beta > 0$, with

$$f(x;\alpha,\beta) = \frac{x^{\alpha-1}(1-x)^{\beta-1}}{B(\alpha,\beta)}, \quad x \in (0,1),$$

where $B(\alpha, \beta)$ denotes the beta function. Let $\boldsymbol{\theta} = (\alpha, \beta)^\top$ be the parameter vector. It can be shown that

$$u_{n\alpha}(\boldsymbol{\theta}) = n[\zeta(\alpha+\beta) - \zeta(\alpha)] + n\log g_1,$$
$$u_{n\beta}(\boldsymbol{\theta}) = n[\zeta(\alpha+\beta) - \zeta(\beta)] + n\log g_2,$$

where $\zeta(\cdot)$ denotes the digamma function, and $g_1 = \prod_{l=1}^n x_l^{1/n}$ and $g_2 = \prod_{l=1}^n (1-x_l)^{1/n}$ are the geometric means of the x_ls and $(1-x_l)$s, respectively. Also,

$$\hat{\boldsymbol{H}}_n(\boldsymbol{\theta}) = \begin{bmatrix} \hat{H}_{\alpha\alpha}(\boldsymbol{\theta}) & \hat{H}_{\alpha\beta}(\boldsymbol{\theta}) \\ \hat{H}_{\beta\alpha}(\boldsymbol{\theta}) & \hat{H}_{\beta\beta}(\boldsymbol{\theta}) \end{bmatrix}, \quad \hat{\boldsymbol{J}}_n(\boldsymbol{\theta}) = \begin{bmatrix} \hat{J}_{\alpha\alpha}(\boldsymbol{\theta}) & \hat{J}_{\alpha\beta}(\boldsymbol{\theta}) \\ \hat{J}_{\beta\alpha}(\boldsymbol{\theta}) & \hat{J}_{\beta\beta}(\boldsymbol{\theta}) \end{bmatrix},$$

$$\hat{\boldsymbol{H}}_n(\boldsymbol{\theta})^{-1} = \begin{bmatrix} \hat{H}^{\alpha\alpha}(\boldsymbol{\theta}) & \hat{H}^{\alpha\beta}(\boldsymbol{\theta}) \\ \hat{H}^{\beta\alpha}(\boldsymbol{\theta}) & \hat{H}^{\beta\beta}(\boldsymbol{\theta}) \end{bmatrix}, \quad \hat{\boldsymbol{J}}_n(\boldsymbol{\theta})^{-1} = \begin{bmatrix} \hat{J}^{\alpha\alpha}(\boldsymbol{\theta}) & \hat{J}^{\alpha\beta}(\boldsymbol{\theta}) \\ \hat{J}^{\beta\alpha}(\boldsymbol{\theta}) & \hat{J}^{\beta\beta}(\boldsymbol{\theta}) \end{bmatrix},$$

whose elements are

$$\hat{H}_{\alpha\alpha}(\boldsymbol{\theta}) = \zeta'(\alpha) - \zeta'(\alpha+\beta), \quad \hat{H}_{\alpha\beta}(\boldsymbol{\theta}) = \hat{H}_{\beta\alpha}(\boldsymbol{\theta}) = -\zeta'(\alpha+\beta),$$
$$\hat{H}_{\beta\beta}(\boldsymbol{\theta}) = \zeta'(\beta) - \zeta'(\alpha+\beta),$$

$$\hat{J}_{\alpha\alpha}(\boldsymbol{\theta}) = [\zeta(\alpha+\beta) - \zeta(\alpha)]^2 + 2[\zeta(\alpha+\beta) - \zeta(\alpha)]\log g_1 + m_1,$$

$$\begin{aligned} \hat{J}_{\alpha\beta}(\boldsymbol{\theta}) = \hat{J}_{\beta\alpha}(\boldsymbol{\theta}) &= [\zeta(\alpha+\beta) - \zeta(\beta)][\zeta(\alpha+\beta) - \zeta(\alpha)] \\ &\quad + [\zeta(\alpha+\beta) - \zeta(\beta)]\log g_1 \\ &\quad + [\zeta(\alpha+\beta) - \zeta(\alpha)]\log g_2 + m_3, \end{aligned}$$

$$\hat{J}_{\beta\beta}(\boldsymbol{\theta}) = [\zeta(\alpha+\beta) - \zeta(\beta)]^2 + 2[\zeta(\alpha+\beta) - \zeta(\beta)]\log g_2 + m_2,$$

$$\begin{aligned} \hat{H}^{\alpha\alpha}(\boldsymbol{\theta}) &= [\hat{H}_{\alpha\alpha}(\boldsymbol{\theta}) - \hat{H}_{\alpha\beta}(\boldsymbol{\theta})^2 \hat{H}_{\beta\beta}(\boldsymbol{\theta})^{-1}]^{-1}, \\ \hat{H}^{\beta\beta}(\boldsymbol{\theta}) &= [\hat{H}_{\beta\beta}(\boldsymbol{\theta}) - \hat{H}_{\alpha\beta}(\boldsymbol{\theta})^2 \hat{H}_{\alpha\alpha}(\boldsymbol{\theta})^{-1}]^{-1}, \\ \hat{H}^{\alpha\beta}(\boldsymbol{\theta}) = \hat{H}^{\beta\alpha}(\boldsymbol{\theta}) &= -\frac{\hat{H}_{\alpha\beta}(\boldsymbol{\theta})\hat{H}_{\alpha\alpha}(\boldsymbol{\theta})^{-1}}{\hat{H}_{\beta\beta}(\boldsymbol{\theta}) - \hat{H}_{\alpha\beta}(\boldsymbol{\theta})^2 \hat{H}_{\alpha\alpha}(\boldsymbol{\theta})^{-1}}, \end{aligned}$$

$$\begin{aligned} \hat{J}^{\alpha\alpha}(\boldsymbol{\theta}) &= [\hat{J}_{\alpha\alpha}(\boldsymbol{\theta}) - \hat{J}_{\alpha\beta}(\boldsymbol{\theta})^2 \hat{J}_{\beta\beta}(\boldsymbol{\theta})^{-1}]^{-1}, \\ \hat{J}^{\beta\beta}(\boldsymbol{\theta}) &= [\hat{J}_{\beta\beta}(\boldsymbol{\theta}) - \hat{J}_{\alpha\beta}(\boldsymbol{\theta})^2 \hat{J}_{\alpha\alpha}(\boldsymbol{\theta})^{-1}]^{-1}, \\ \hat{J}^{\alpha\beta}(\boldsymbol{\theta}) = \hat{J}^{\beta\alpha}(\boldsymbol{\theta}) &= -\frac{\hat{J}_{\alpha\beta}(\boldsymbol{\theta})\hat{J}_{\alpha\alpha}(\boldsymbol{\theta})^{-1}}{\hat{J}_{\beta\beta}(\boldsymbol{\theta}) - \hat{J}_{\alpha\beta}(\boldsymbol{\theta})^2 \hat{J}_{\alpha\alpha}(\boldsymbol{\theta})^{-1}}, \end{aligned}$$

$$m_1 = \frac{1}{n}\sum_{l=1}^{n}[\log x_l]^2, \quad m_2 = \frac{1}{n}\sum_{l=1}^{n}[\log(1-x_l)]^2,$$
$$m_3 = \frac{1}{n}\sum_{l=1}^{n}\log(1-x_l)\log x_l,$$

and $\zeta'(\cdot)$ denotes the trigamma function.

Let $\hat{\boldsymbol{\theta}} = (\hat{\alpha}, \hat{\beta})^\top$ be the unrestricted MLE of $\boldsymbol{\theta} = (\alpha, \beta)^\top$. First, suppose the interest lies in testing the null hypothesis $\mathcal{H}_0 : \alpha = \alpha_0$ against $\mathcal{H}_a : \alpha \neq \alpha_0$, where β acts as a nuisance parameter. Let $\tilde{\boldsymbol{\theta}} = (\alpha_0, \tilde{\beta})^\top$ denote the restricted MLE of $\boldsymbol{\theta} = (\alpha, \beta)^\top$ obtained under the null hypothesis $\mathcal{H}_0 : \alpha = \alpha_0$. The statistic T^* assumes the form

$$T^* = n(\hat{\alpha} - \alpha_0)[\zeta(\alpha_0 + \tilde{\beta}) - \zeta(\alpha_0) + \log g_1][\hat{\boldsymbol{J}}_n(\tilde{\boldsymbol{\theta}})^{-1}\hat{\boldsymbol{H}}_n(\tilde{\boldsymbol{\theta}})]_{\alpha\alpha}.$$

Now, let $\mathcal{H}_0 : \beta = \beta_0$ be the null hypothesis of interest, and $\tilde{\boldsymbol{\theta}} = (\tilde{\alpha}, \beta_0)^\top$ be the restricted MLE of $\boldsymbol{\theta} = (\alpha, \beta)^\top$ obtained under the null hypothesis $\mathcal{H}_0 : \beta = \beta_0$, where α acts as a nuisance parameter. For testing this null hypothesis, the robust gradient statistic becomes

$$T^* = n(\hat{\beta} - \beta_0)[\zeta(\tilde{\alpha} + \beta_0) - \zeta(\beta_0) + \log g_2][\hat{\boldsymbol{J}}_n(\tilde{\boldsymbol{\theta}})^{-1}\hat{\boldsymbol{H}}_n(\tilde{\boldsymbol{\theta}})]_{\beta\beta}.$$

Example 4.5 (Exponential-Poisson distribution). Let $x_1, \ldots, x_n$ be a sample of size n from a gamma distribution with shape parameter $\alpha > 0$ and scale parameter $\xi > 0$. We shall assume that the data come from an exponential-Poisson (EP) distribution proposed by Kuş [43]. The EP probability density function is given by

$$f(x; \lambda, \beta) = \frac{\lambda\beta}{e^{\lambda} - 1} \exp\left(-\beta x + \lambda e^{-\beta x}\right), \quad x > 0,$$

where $\lambda > 0$ and $\beta > 0$. This model was obtained under the concept of population heterogeneity (through the process of compounding). For all values of parameters, the EP density is strictly decreasing in x and tends to zero as $x \to \infty$. As λ approaches zero, the EP leads to exponential distribution with parameter β. Let $\boldsymbol{\theta} = (\lambda, \beta)^\top$ be the parameter vector. We have that

$$u_{n\lambda}(\boldsymbol{\theta}) = n[\lambda^{-1} - (1 - e^{-\lambda})^{-1} + \bar{x}_1], \quad u_{n\beta}(\boldsymbol{\theta}) = n[\beta^{-1} - \bar{x} - \lambda\bar{x}_2],$$

where $\bar{x} = n^{-1}\sum_{l=1}^{n} x_l$,

$$\bar{x}_1 = \frac{1}{n}\sum_{l=1}^{n} e^{-\beta x_l}, \quad \bar{x}_2 = \frac{1}{n}\sum_{l=1}^{n} x_l e^{-\beta x_l}.$$

Also,

$$\hat{\boldsymbol{H}}_n(\boldsymbol{\theta}) = \begin{bmatrix} \hat{H}_{\lambda\lambda}(\boldsymbol{\theta}) & \hat{H}_{\lambda\beta}(\boldsymbol{\theta}) \\ \hat{H}_{\beta\lambda}(\boldsymbol{\theta}) & \hat{H}_{\beta\beta}(\boldsymbol{\theta}) \end{bmatrix}, \quad \hat{\boldsymbol{J}}_n(\boldsymbol{\theta}) = \begin{bmatrix} \hat{J}_{\lambda\lambda}(\boldsymbol{\theta}) & \hat{J}_{\lambda\beta}(\boldsymbol{\theta}) \\ \hat{J}_{\beta\lambda}(\boldsymbol{\theta}) & \hat{J}_{\beta\beta}(\boldsymbol{\theta}) \end{bmatrix},$$

$$\hat{\boldsymbol{H}}_n(\boldsymbol{\theta})^{-1} = \begin{bmatrix} \hat{H}^{\lambda\lambda}(\boldsymbol{\theta}) & \hat{H}^{\lambda\beta}(\boldsymbol{\theta}) \\ \hat{H}^{\beta\lambda}(\boldsymbol{\theta}) & \hat{H}^{\beta\beta}(\boldsymbol{\theta}) \end{bmatrix}, \quad \hat{\boldsymbol{J}}_n(\boldsymbol{\theta})^{-1} = \begin{bmatrix} \hat{J}^{\lambda\lambda}(\boldsymbol{\theta}) & \hat{J}^{\lambda\beta}(\boldsymbol{\theta}) \\ \hat{J}^{\beta\lambda}(\boldsymbol{\theta}) & \hat{J}^{\beta\beta}(\boldsymbol{\theta}) \end{bmatrix},$$

where

$$\hat{H}_{\lambda\lambda}(\boldsymbol{\theta}) = \frac{1 + e^{2\lambda} - \lambda^2 e^{\lambda} - 2e^{\lambda}}{\lambda^2(1 - e^{-\lambda})^2}, \quad \hat{H}_{\lambda\beta}(\boldsymbol{\theta}) = \hat{H}_{\beta\lambda}(\boldsymbol{\theta}) = \bar{x}_2,$$
$$\hat{H}_{\beta\beta}(\boldsymbol{\theta}) = \beta^{-2} - \lambda m_4,$$
$$\hat{J}_{\lambda\lambda}(\boldsymbol{\theta}) = [\lambda^{-1} - (1 - e^{-\lambda})^{-1}]^2 + 2[\lambda^{-1} - (1 - e^{-\lambda})^{-1}]\bar{x}_1 + m_1,$$

$$\hat{J}_{\lambda\beta}(\boldsymbol{\theta}) = \hat{J}_{\beta\lambda}(\boldsymbol{\theta}) = (\beta^{-1} - \bar{x})[\lambda^{-1} - (1 - e^{-\lambda})^{-1}] - \beta^{-1}\bar{x}_1$$
$$- \{\lambda[\lambda^{-1} - (1 - e^{-\lambda})^{-1}] + 1\}\bar{x}_2 - \lambda m_2,$$

$$\hat{J}_{\beta\beta}(\boldsymbol{\theta}) = \beta^{-1}(\beta^{-1} - \bar{x} - \lambda\bar{x}_2) - \beta^{-1}\bar{x} + m_3$$
$$- \lambda\beta^{-1}\bar{x}_2 + 2\lambda m_4 + \lambda^2 m_5,$$

$$\hat{H}^{\lambda\lambda}(\boldsymbol{\theta}) = [\hat{H}_{\lambda\lambda}(\boldsymbol{\theta}) - \hat{H}_{\lambda\beta}(\boldsymbol{\theta})^2\hat{H}_{\beta\beta}(\boldsymbol{\theta})^{-1}]^{-1},$$
$$\hat{H}^{\beta\beta}(\boldsymbol{\theta}) = [\hat{H}_{\beta\beta}(\boldsymbol{\theta}) - \hat{H}_{\lambda\beta}(\boldsymbol{\theta})^2\hat{H}_{\lambda\lambda}(\boldsymbol{\theta})^{-1}]^{-1},$$
$$\hat{H}^{\lambda\beta}(\boldsymbol{\theta}) = \hat{H}^{\beta\lambda}(\boldsymbol{\theta}) = -\frac{\hat{H}_{\lambda\beta}(\boldsymbol{\theta})\hat{H}_{\lambda\lambda}(\boldsymbol{\theta})^{-1}}{\hat{H}_{\beta\beta}(\boldsymbol{\theta}) - \hat{H}_{\lambda\beta}(\boldsymbol{\theta})^2\hat{H}_{\lambda\lambda}(\boldsymbol{\theta})^{-1}},$$

$$\hat{J}^{\lambda\lambda}(\boldsymbol{\theta}) = [\hat{J}_{\lambda\lambda}(\boldsymbol{\theta}) - \hat{J}_{\lambda\beta}(\boldsymbol{\theta})^2\hat{J}_{\beta\beta}(\boldsymbol{\theta})^{-1}]^{-1},$$
$$\hat{J}^{\beta\beta}(\boldsymbol{\theta}) = [\hat{J}_{\beta\beta}(\boldsymbol{\theta}) - \hat{J}_{\lambda\beta}(\boldsymbol{\theta})^2\hat{J}_{\lambda\lambda}(\boldsymbol{\theta})^{-1}]^{-1},$$
$$\hat{J}^{\lambda\beta}(\boldsymbol{\theta}) = \hat{J}^{\beta\lambda}(\boldsymbol{\theta}) = -\frac{\hat{J}_{\lambda\beta}(\boldsymbol{\theta})\hat{J}_{\lambda\lambda}(\boldsymbol{\theta})^{-1}}{\hat{J}_{\beta\beta}(\boldsymbol{\theta}) - \hat{J}_{\lambda\beta}(\boldsymbol{\theta})^2\hat{J}_{\lambda\lambda}(\boldsymbol{\theta})^{-1}},$$

$$m_1 = \frac{1}{n}\sum_{l=1}^{n} e^{-2\beta x_l}, \quad m_2 = \frac{1}{n}\sum_{l=1}^{n} x_l e^{-2\beta x_l},$$
$$m_3 = \frac{1}{n}\sum_{l=1}^{n} x_l^2, \quad m_4 = \frac{1}{n}\sum_{l=1}^{n} x_l^2 e^{-\beta x_l}, \quad m_5 = \frac{1}{n}\sum_{l=1}^{n} x_l^2 e^{-2\beta x_l}.$$

Let $\hat{\boldsymbol{\theta}} = (\hat{\lambda}, \hat{\beta})^\top$ be the unrestricted MLE of $\boldsymbol{\theta} = (\lambda, \beta)^\top$. We wish to test the null hypothesis $\mathcal{H}_0 : \lambda = \lambda_0$ against $\mathcal{H}_a : \lambda \neq \lambda_0$, where β acts as a nuisance parameter. Let $\tilde{\boldsymbol{\theta}} = (\lambda_0, \tilde{\beta})^\top$ denote the restricted MLE of $\boldsymbol{\theta} = (\lambda, \beta)^\top$ obtained under the null hypothesis $\mathcal{H}_0 : \lambda = \lambda_0$. The statistic T^* takes the form

$$T^* = n(\hat{\lambda} - \lambda_0)[\lambda_0^{-1} - (1 - e^{-\lambda_0})^{-1} + \tilde{\bar{x}}_1][\hat{\boldsymbol{J}}_n(\tilde{\boldsymbol{\theta}})^{-1}\hat{\boldsymbol{H}}_n(\tilde{\boldsymbol{\theta}})]_{\lambda\lambda},$$

where $\tilde{\bar{x}}_1 = n^{-1}\sum_{l=1}^{n} e^{-\tilde{\beta}x_l}$. Finally, suppose we want to test the null hypothesis $\mathcal{H}_0 : \beta = \beta_0$. Let $\tilde{\boldsymbol{\theta}} = (\tilde{\lambda}, \beta_0)^\top$ be the restricted MLE of $\boldsymbol{\theta} = (\lambda, \beta)^\top$ obtained under the null hypothesis $\mathcal{H}_0 : \beta = \beta_0$, where λ acts as a nuisance parameter. Hence, we have that

$$T^* = n(\hat{\beta} - \beta_0)[\beta_0^{-1} - \bar{x} - \tilde{\lambda}\tilde{\bar{x}}_2][\hat{\boldsymbol{J}}_n(\tilde{\boldsymbol{\theta}})^{-1}\hat{\boldsymbol{H}}_n(\tilde{\boldsymbol{\theta}})]_{\beta\beta},$$

where $\tilde{\bar{x}}_2 = n^{-1}\sum_{l=1}^{n} x_l e^{-\beta_0 x_l}$.

4.4 NUMERICAL RESULTS

In this section, we present Monte Carlo simulation experiments in order to verify the robustness of the robust gradient statistic under model misspecification. All simulations were carried out using the Ox matrix programming language (http://www.doornik.com). The number of Monte Carlo replications was 20,000. For the sake of comparison, we also consider in the numerical study the usual likelihood ratio, Wald and score tests, and the robust versions of the Wald and score tests.

First, we consider a special case of Example (4.1), where a sample of size $n = 115$ is obtained from a Weibull distribution with shape parameter $a = 2$ and scale parameter $b = 1$, but we will assume that the data come from a gamma model with shape parameter $\alpha = \Gamma(1.5)$ and scale parameter θ. In this setup, we want to test the null hypothesis $\mathcal{H}_0 : \theta = 1$ against $\mathcal{H}_a : \theta \neq 1$. We report the null rejection rates (ie, the percentage of times that the corresponding statistics exceed the 10%, 5%, and 1% upper points of the reference χ_1^2 distribution) of the tests at the 10%, 5%, and 1% nominal significance levels under model correctness and under model misspecification. The results are presented in Table 4.1 and the entries are percentages. As expected, the tests which are based on the statistics S_{LR}, S_{W}, S_{R}, and S_{T} present a good agreement with the true significance level under model correctness. On the other hand, these tests are much size distorted under model misspecification, whereas the robust tests present a very good performance.

We now present a brief study of power properties of the robust tests for the special case of Example (4.1) considered above. We set $n = 200$ and 500, and $\gamma = 5\%$. For the power simulations we compute the rejection rates under the alternative hypothesis $\mathcal{H}_a : \theta = \delta$, for different values of δ. The

Table 4.1 Null Rejection Rates (%) for $\mathcal{H}_0 : \theta = 1$ With $n = 115$

	Under Model Correctness		
Statistic	$\gamma = 10\%$	$\gamma = 5\%$	$\gamma = 1\%$
S_{LR}	9.86	5.00	1.03
S_W	10.16	5.41	1.54
S_R	9.80	4.93	1.06
S_T	9.80	4.93	1.06
W^*	8.69	5.14	2.02
R^*	10.62	5.67	1.62
T^*	9.68	4.71	0.90
	Under Model Misspecification		
Statistic	$\gamma = 10\%$	$\gamma = 5\%$	$\gamma = 1\%$
S_{LR}	0.09	0.00	0.00
S_W	0.19	0.03	0.00
S_R	0.07	0.01	0.00
S_T	0.07	0.01	0.00
W^*	9.62	4.86	1.16
R^*	10.15	4.87	1.01
T^*	10.09	4.68	0.85

power of the robust tests are displayed in Fig. 4.1. As can be seen from this figure, none of the robust tests is uniformly most powerful for testing the null hypothesis $\mathcal{H}_0 : \theta = 1$. Additionally, as the sample size grows, the power curves of the robust tests become indistinguishable from each other, as expected.

Next, we take into account the situation presented in Example (4.2). We set the sample size at $n = 85$, $\mu = 1$, $\beta = 1$, $\eta = 1$ and we want to test the null hypothesis $\mathcal{H}_0 : \alpha = 1$. Fig. 4.2A displays curves of quantile relative discrepancies versus the correspondent asymptotic quantiles for the statistics S_{LR}, S_W, S_R, and S_T, whereas Fig. 4.2B displays curves of quantile relative discrepancies versus the correspondent asymptotic quantiles for the statistics W^*, R^*, and T^*. Relative quantile discrepancy is defined as the difference between exact (estimated by simulation) and asymptotic quantiles divided by the latter. The closer to zero the relative quantile discrepancy, the better is the approximation of the exact null distribution (ie, the exact distribution under the null hypothesis) of the test statistic by the limiting χ^2 distribution. Fig. 4.2A reveals that the distributions of the statistics S_{LR},

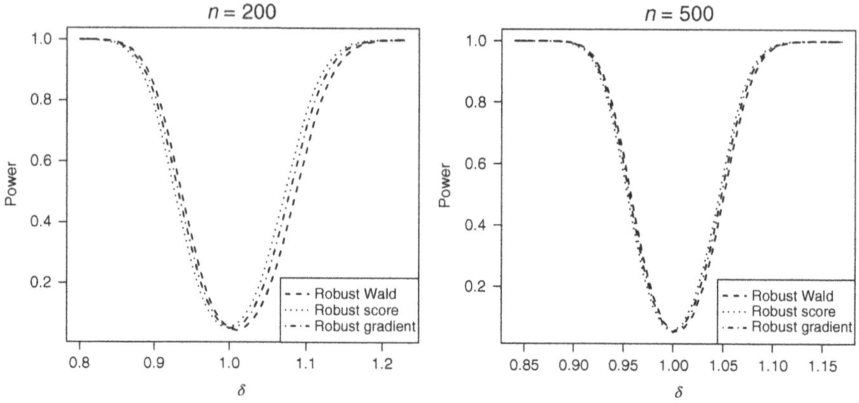

Fig. 4.1 Power of the robust tests for $\gamma = 5\%$.

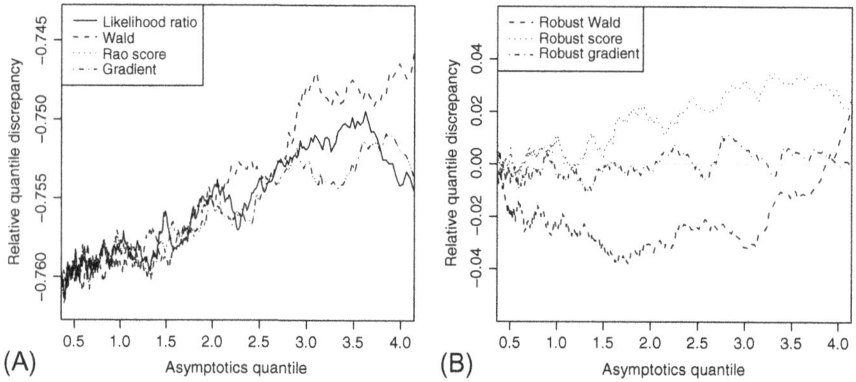

Fig. 4.2 Quantile relative discrepancies: (A) usual statistics; (B) robust statistics.

S_{W}, S_{R}, and S_{T} are poorly approximated by the reference χ_1^2 distribution under model misspecification. On the other hand, the distributions of the robust statistics W^*, R^*, and T^* are close to the reference χ_1^2 distribution (see Fig. 4.2B).

Now, we take into account Example (4.3). We consider $\theta = 0$ and want to test the null hypothesis $\mathcal{H}_0 : \mu = 0$. We report the null rejection rates for $\mathcal{H}_0$ of the tests (usual and robust tests) at the 10% and 5% nominal significance levels by considering different sample sizes. We set $n = 25, 50, 90, 120, 150,$ and 200. The results are presented in Table 4.2 and the entries are percentages. It is evident that the usual tests are markedly liberal (rejecting the null hypothesis more frequently than expected based on the selected nominal

Table 4.2 Null Rejection Rates (%) for $\mathcal{H}_0 : \mu = 0$

	$n = 25$		$n = 50$	
Statistic	$\gamma = 10\%$	$\gamma = 5\%$	$\gamma = 10\%$	$\gamma = 5\%$
S_{LR}	20.09	12.80	20.57	13.15
S_W	18.01	11.14	18.03	10.94
S_R	22.60	15.37	22.84	15.49
S_T	20.30	12.97	20.60	13.30
W^*	11.42	5.77	10.35	5.32
R^*	9.91	4.02	9.45	4.57
T^*	10.64	4.95	9.94	5.01
	$n = 90$		$n = 120$	
Statistic	$\gamma = 10\%$	$\gamma = 5\%$	$\gamma = 10\%$	$\gamma = 5\%$
S_{LR}	20.82	13.29	20.91	13.77
S_W	18.33	11.16	18.48	11.62
S_R	23.20	15.74	23.39	15.93
S_T	20.84	13.37	20.92	13.79
W^*	10.37	5.05	10.03	5.15
R^*	9.89	4.63	9.62	4.86
T^*	10.11	4.82	9.77	5.00
	$n = 150$		$n = 200$	
Statistic	$\gamma = 10\%$	$\gamma = 5\%$	$\gamma = 10\%$	$\gamma = 5\%$
S_{LR}	21.16	13.80	21.13	13.74
S_W	18.63	11.57	18.75	11.51
S_R	23.77	16.00	23.54	16.02
S_T	21.18	13.83	21.13	13.77
W^*	10.24	4.91	9.69	4.91
R^*	9.93	4.67	9.44	4.70
T^*	10.11	4.80	9.55	4.80

level) if the working model assumption is incorrect for the sample sizes considered. It means that the asymptotic χ^2 distribution of the usual statistics no longer holds under model misspecification. On the other hand, the robust tests present a good agreement with the true significance levels under model misspecification for the sample sizes considered; that is, these tests perform remarkably well if the working model assumption is incorrect. In short, the inference based on the robust test statistics is much more accurate than that based on the usual test statistics if the working model assumption is incorrect.

Finally, suppose we have a sample $x_1, \ldots, x_n$ of size n from the Student-t distribution with location parameter $\theta \in \mathbb{R}$ and degrees of freedom $\nu > 0$. We shall assume that the data are from a normal distribution with mean $\mu \in \mathbb{R}$ and known variance $\sigma^2 > 0$, which is assumed to be equal to one for simplicity. As noted earlier, if the density $g(x)$ (the true model) is a member of the parametric family $\{f(x; \boldsymbol{\theta}); \boldsymbol{\theta} \in \boldsymbol{\Theta}\}$, then the usual test statistics might be used. The type of model misspecification we have considered in this chapter means that the density $g(x)$ does not belong to the parametric family $\{f(x; \boldsymbol{\theta}); \boldsymbol{\theta} \in \boldsymbol{\Theta}\}$. In this case we have that the Student-t distribution reduces to the normal distribution when ν is sufficiently large. We shall provide some Monte Carlo simulations to show that even in this case the usual tests do not present a good performance in testing the null hypothesis $\mathcal{H}_0 : \mu = \mu_0$. For testing this null hypothesis, the usual gradient statistic is given by

$$S_{\mathrm{T}} = n(\bar{x} - \mu_0)^2,$$

where $\bar{x} = n^{-1}\sum_{l=1}^{n} x_l$. The robust gradient test statistic can be expressed as

$$T^* = S_{\mathrm{T}} \left\{ \frac{1}{n} \sum_{l=1}^{n} (x_l - \bar{x})^2 + (\bar{x} - \mu_0)^2 \right\}^{-1}.$$

The quantity between curly braces can be considered as a kind of "correction factor" for the usual gradient statistic, which makes this statistic to have asymptotic χ_1^2 distribution under the null hypothesis and under model misspecification. It can be shown that $S_{\mathrm{LR}} = S_{\mathrm{W}} = S_{\mathrm{R}} = S_{\mathrm{T}}$ and $W^* = R^* = T^*$. Now, we set the sample size at $n = 60$, $\theta = 0$, and $\nu = 3$, 8, 15, and 50. We wish to test the null hypothesis $\mathcal{H}_0 : \mu = 0$. Fig. 4.3 displays quantile-quantile plots, where the exact quantiles (estimated by Monte Carlo simulations) of the usual test statistics (*dash*) and robust test statistics (*dot*) are plotted against their asymptotic quantiles for different degrees of freedom. Note that the quantiles of the usual statistics exceed their asymptotic counterparts, especially when the value of the degrees of freedom is small. It is also noteworthy that the exact quantiles of the robust test statistics are in agreement with the corresponding asymptotic quantiles in all cases, thus revealing that its null distribution is well approximated by the limiting null distribution used in the test, whereas the usual test statistics need a large value for the degrees of freedom.

It should be mentioned that much more numerical work is needed to come to any general conclusions about the performance of the robust Wald, score, and gradient statistics for testing hypotheses under model

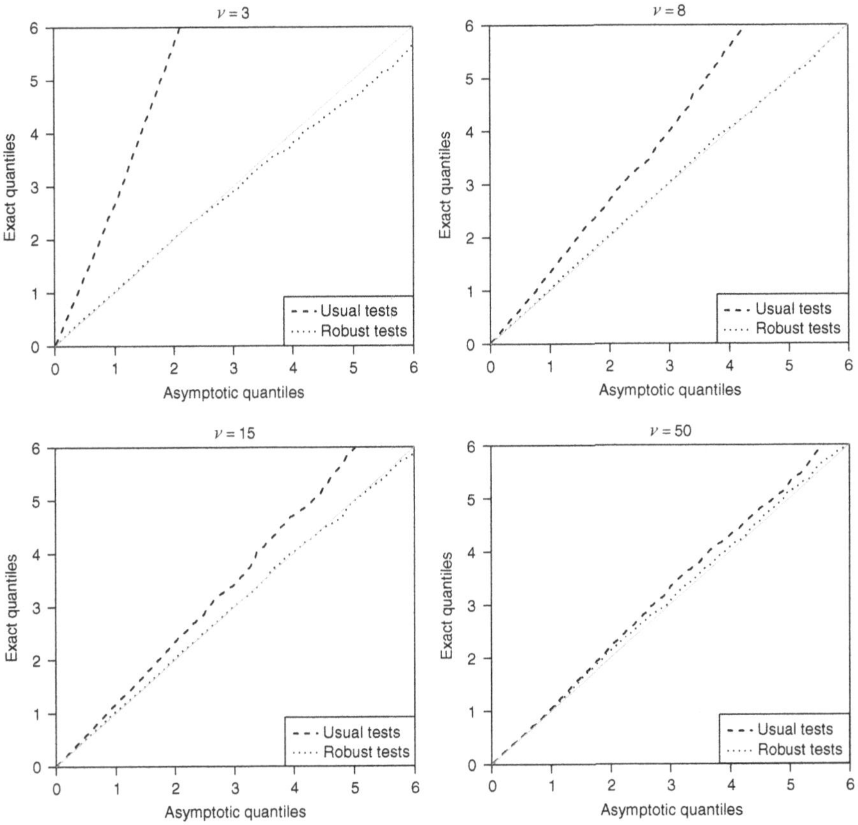

Fig. 4.3 Exact versus asymptotic quantiles of the test statistics.

misspecification as well as recommend one of them. Future research regarding Monte Carlo simulations on hypothesis testing inference under model misspecification in several parametric models can be considered. From these numerical results, one can recommend one of them (robust Wald, score, and gradient statistics) to test hypotheses under model misspecification at least in specific classes of parametric models.

CHAPTER 5

The Robust Gradient-Type Bounded-Influence Test

5.1 INTRODUCTION

Let $\{F_{\boldsymbol{\theta}}, \boldsymbol{\theta} \in \boldsymbol{\Theta}\}$ be a general parametric model, where $\boldsymbol{\Theta}$ is an open convex subset of $\mathbb{R}^p$ known as the parameter space. Let $f_{\boldsymbol{\theta}}$ be the corresponding probability density function of $F_{\boldsymbol{\theta}}$. Let $\boldsymbol{\theta} = (\boldsymbol{\theta}_{(1)}^\top, \boldsymbol{\theta}_{(2)}^\top)^\top$, where $\boldsymbol{\theta}_{(1)} = (\theta_1, \ldots, \theta_{p-q})^\top$ and $\boldsymbol{\theta}_{(2)} = (\theta_{p-q+1}, \ldots, \theta_p)^\top$ are vectors of dimensions $p - q$ and q, respectively. We are interested in testing the composite null hypothesis $\mathcal{H}_0 : \boldsymbol{\theta}_{(2)} = \boldsymbol{\theta}_{0(2)}$ against the two-sided alternative hypothesis $\mathcal{H}_{\mathrm{a}} : \boldsymbol{\theta}_{(2)} \neq \boldsymbol{\theta}_{0(2)}$. From now on, we assume that $\boldsymbol{\theta}_{0(2)} = \mathbf{0}_q$.

Let $z_1, \ldots, z_n$ be n independent and identically distributed random variables. The proposed test relies on M-estimators $\boldsymbol{T}_n = (\boldsymbol{T}_{n(1)}^\top, \boldsymbol{T}_{n(2)}^\top)^\top$ of $\boldsymbol{\theta} = (\boldsymbol{\theta}_{(1)}^\top, \boldsymbol{\theta}_{(2)}^\top)^\top$, defined implicitly as the solution in $\boldsymbol{\theta}$ of

$$\sum_{l=1}^{n} \boldsymbol{\psi}(z_l, \boldsymbol{\theta}) = \mathbf{0}_p, \tag{5.1}$$

where the dimension of $\boldsymbol{\psi}(z, \boldsymbol{\theta})$ is the same as that of $\boldsymbol{\theta}$. It is well known that choosing a bounded $\boldsymbol{\psi}$-function or controlling the bound on $\boldsymbol{\psi}$ defines a robust estimator. For example, a simple choice for $\boldsymbol{\psi}$ is given by a weighted score function leading to a weighted MLE, with smaller weights when the score function becomes too large, that is,

$$\boldsymbol{\psi}_c(z, \boldsymbol{\theta}) = w_c(z, \boldsymbol{\theta}) \boldsymbol{s}(z, \boldsymbol{\theta}),$$

The Gradient Test. http://dx.doi.org/10.1016/B978-0-12-803596-2.00005-3

where $\boldsymbol{s}(z,\boldsymbol{\theta}) = \partial \log f_{\boldsymbol{\theta}}(z)/\partial\boldsymbol{\theta}$ is the score function, and the weight function can be defined through the Huber function with parameter $c > 0$ given by

$$w_c(z,\boldsymbol{\theta}) = \min\left\{1, \frac{c}{\|\boldsymbol{s}(z,\boldsymbol{\theta})\|}\right\},$$

where $\|\cdot\|$ denotes the Euclidean norm.

Under quite general conditions, M-estimators are consistent at the model and asymptotically normally distributed (see, eg, [44, 45]). The influence function (IF) and asymptotic covariance matrix of M-estimators are

$$\begin{aligned}\mathrm{IF}(z;\boldsymbol{\psi},F_{\boldsymbol{\theta}}) &= \boldsymbol{M}(\boldsymbol{\psi},F_{\boldsymbol{\theta}})^{-1}\boldsymbol{\psi}(z,\boldsymbol{\theta}),\\ \boldsymbol{V}(\boldsymbol{\psi},F_{\boldsymbol{\theta}}) &= \boldsymbol{M}(\boldsymbol{\psi},F_{\boldsymbol{\theta}})^{-1}\boldsymbol{Q}(\boldsymbol{\psi},F_{\boldsymbol{\theta}})\boldsymbol{M}(\boldsymbol{\psi},F_{\boldsymbol{\theta}})^{-\top},\end{aligned}$$

respectively, where

$$\boldsymbol{M}(\boldsymbol{\psi},F_{\boldsymbol{\theta}}) = -\int \frac{\partial\boldsymbol{\psi}(z,\boldsymbol{\theta})}{\partial\boldsymbol{\theta}}\,dF_{\boldsymbol{\theta}}(z),$$

and

$$\boldsymbol{Q}(\boldsymbol{\psi},F_{\boldsymbol{\theta}}) = \int \boldsymbol{\psi}(z,\boldsymbol{\theta})\boldsymbol{\psi}(z,\boldsymbol{\theta})^{\top}\,dF_{\boldsymbol{\theta}}(z).$$

From the partition of $\boldsymbol{\theta}$, we have the corresponding partitions:

$$\begin{aligned}\boldsymbol{\psi}(z,\boldsymbol{\theta}) &= (\boldsymbol{\psi}(z,\boldsymbol{\theta})_{(1)}^{\top}, \boldsymbol{\psi}(z,\boldsymbol{\theta})_{(2)}^{\top})^{\top},\\ \boldsymbol{V}(\boldsymbol{\psi},F_{\boldsymbol{\theta}}) &= \begin{bmatrix}\boldsymbol{V}(\boldsymbol{\psi},F_{\boldsymbol{\theta}})_{(11)} & \boldsymbol{V}(\boldsymbol{\psi},F_{\boldsymbol{\theta}})_{(12)}\\ \boldsymbol{V}(\boldsymbol{\psi},F_{\boldsymbol{\theta}})_{(21)} & \boldsymbol{V}(\boldsymbol{\psi},F_{\boldsymbol{\theta}})_{(22)}\end{bmatrix},\\ \boldsymbol{M}(\boldsymbol{\psi},F_{\boldsymbol{\theta}}) &= \begin{bmatrix}\boldsymbol{M}(\boldsymbol{\psi},F_{\boldsymbol{\theta}})_{(11)} & \boldsymbol{M}(\boldsymbol{\psi},F_{\boldsymbol{\theta}})_{(12)}\\ \boldsymbol{M}(\boldsymbol{\psi},F_{\boldsymbol{\theta}})_{(21)} & \boldsymbol{M}(\boldsymbol{\psi},F_{\boldsymbol{\theta}})_{(22)}\end{bmatrix},\\ \boldsymbol{Q}(\boldsymbol{\psi},F_{\boldsymbol{\theta}}) &= \begin{bmatrix}\boldsymbol{Q}(\boldsymbol{\psi},F_{\boldsymbol{\theta}})_{(11)} & \boldsymbol{Q}(\boldsymbol{\psi},F_{\boldsymbol{\theta}})_{(12)}\\ \boldsymbol{Q}(\boldsymbol{\psi},F_{\boldsymbol{\theta}})_{(21)} & \boldsymbol{Q}(\boldsymbol{\psi},F_{\boldsymbol{\theta}})_{(22)}\end{bmatrix}.\end{aligned}$$

Heritier and Ronchetti [41] introduced three classes of tests in general parametric models: (i) the Wald-type test; (ii) the score-type test; and (iii) the likelihood ratio-type test. The Wald-type test statistic is defined as

$$W_n^2 = \boldsymbol{T}_{n(2)}^{\top}\boldsymbol{V}(\boldsymbol{\psi},F_{\boldsymbol{\theta}})_{(22)}^{-1}\boldsymbol{T}_{n(2)},$$

where the matrix $\boldsymbol{V}(\boldsymbol{\psi},F_{\boldsymbol{\theta}})_{(22)}$ can be estimated by replacing $\boldsymbol{\theta}$ with $\boldsymbol{T}_n$ in $\boldsymbol{V}(\boldsymbol{\psi},F_{\boldsymbol{\theta}})$. The score-type test statistic is given by

$$R_n^2 = \boldsymbol{Z}_n^{\top}\boldsymbol{C}(\boldsymbol{\psi},F_{\boldsymbol{\theta}})^{-1}\boldsymbol{Z}_n,$$

where

$$Z_n = \frac{1}{n}\sum_{l=1}^{n} \psi(z_l, T_n^{\omega})_{(2)},$$

$C(\psi, F_\theta) = M(\psi, F_\theta)_{(22.1)} V(\psi, F_\theta)_{(22)} M(\psi, F_\theta)_{(22.1)}^{\top}$ is a $q \times q$ positive definite matrix, $M(\psi, F_\theta)_{(22.1)}$ assumes the form $M(\psi, F_\theta)_{(22.1)} = M(\psi, F_\theta)_{(22)} - M(\psi, F_\theta)_{(21)} M(\psi, F_\theta)_{(11)}^{-1} M(\psi, F_\theta)_{(12)}$, and $T_n^{\omega} = (T_{n(1)}^{\omega\top}, T_{n(2)}^{\omega\top})^{\top}$ is the M-estimator in the reduced model, which can be obtained from the solution of the equation

$$\sum_{l=1}^{n} \psi(z_l, T_n^{\omega})_{(1)} = \mathbf{0}_{p-q} \quad \text{with} \quad T_{n(2)}^{\omega} = \theta_{0(2)}. \tag{5.2}$$

The matrix $C(\psi, F_\theta)$ is the asymptotic covariance matrix of Z_n, and it can be estimated consistently by replacing θ with T_n^{ω}. The likelihood ratio-type test statistic takes the form

$$S_n^2 = \frac{2}{n}\sum_{l=1}^{n}[\rho(z_l, T_n) - \rho(z_l, T_n^{\omega})],$$

where $(\partial\rho/\partial\theta)(z, \theta) = \psi(z, \theta)$, and T_n and T_n^{ω} are the M-estimators of θ in the full and reduced models, defined by Eqs. (5.1) and (5.2), respectively. Heritier and Ronchetti [41] proved that the Wald- and score-type test statistics have asymptotically a central χ_q^2 distribution under the null hypothesis, where $q \leq p$ is the dimension of the hypothesis to be tested, and a noncentral $\chi_{q,\lambda}^2$ distribution under a sequence of contiguous alternatives, with the same noncentrality parameter λ. On the other hand, the asymptotic distribution of the likelihood ratio-type test statistic is a linear combination of random variables having χ_1^2 distribution. Therefore, the robust Wald and score statistics have the same asymptotic distributional properties under either the null or contiguous alternatives hypotheses as their classical counterpart, whereas the likelihood ratio-type statistic has a much more complicated asymptotic distribution.

In this chapter, we propose a new robust statistic that arises from the Wald- and score-type test statistics. As in the classical case, the score-type statistic measures the squared length of Z_n evaluated at the null hypothesis $\mathcal{H}_0$ using the metric given by the inverse of the matrix $C(\psi, F_\theta)$, whereas the Wald type statistic gives the squared of the M estimator $T_{n(2)}$ using the metric given by the inverse of the matrix $V(\psi, F_\theta)_{(22)}$. Moreover, both are quadratic forms. In other words, the Wald- and score-type statistics can be expressed as

$$W_n^2 = \|\boldsymbol{T}_{n(2)}\|^2_{\boldsymbol{V}(\boldsymbol{\psi},F_{\boldsymbol{\theta}})^{-1}_{(22)}}, \quad R_n^2 = \|\boldsymbol{Z}_n\|^2_{\boldsymbol{C}(\boldsymbol{\psi},F_{\boldsymbol{\theta}})^{-1}},$$

where $\|\boldsymbol{x}\|^2_{\boldsymbol{\Sigma}} = \boldsymbol{x}^\top \boldsymbol{\Sigma}\boldsymbol{x}$. The proposed robust statistic, on the other hand, is not a quadratic form, and it is the natural counterpart of the corresponding classical gradient test statistic.

5.2 THE ROBUST GRADIENT-TYPE TEST

The new robust statistic is proposed on the basis of the Wald- and score-type test statistics. Note that we can express the statistics W_n^2 and R_n^2 in the forms

$$\begin{aligned} W_n^2 &= \boldsymbol{T}_{n(2)}^\top \boldsymbol{V}(\boldsymbol{\psi},F_{\boldsymbol{\theta}})^{-1/2}_{(22)} \boldsymbol{V}(\boldsymbol{\psi},F_{\boldsymbol{\theta}})^{-1/2}_{(22)} \boldsymbol{T}_{n(2)} \\ &= [\boldsymbol{V}(\boldsymbol{\psi},F_{\boldsymbol{\theta}})^{-1/2}_{(22)} \boldsymbol{T}_{n(2)}]^\top \boldsymbol{V}(\boldsymbol{\psi},F_{\boldsymbol{\theta}})^{-1/2}_{(22)} \boldsymbol{T}_{n(2)}, \\ R_n^2 &= \boldsymbol{Z}_n^\top \boldsymbol{C}(\boldsymbol{\psi},F_{\boldsymbol{\theta}})^{-1/2} \boldsymbol{C}(\boldsymbol{\psi},F_{\boldsymbol{\theta}})^{-1/2} \boldsymbol{Z}_n \\ &= [\boldsymbol{C}(\boldsymbol{\psi},F_{\boldsymbol{\theta}})^{-1/2} \boldsymbol{Z}_n]^\top \boldsymbol{C}(\boldsymbol{\psi},F_{\boldsymbol{\theta}})^{-1/2} \boldsymbol{Z}_n, \end{aligned}$$

where $\boldsymbol{A}^{1/2}$ stands for any root of the matrix $\boldsymbol{A}$, that is, any solution of $(\boldsymbol{A}^{1/2})^\top \boldsymbol{A}^{1/2} = \boldsymbol{A}$, and $\boldsymbol{A}^{-1/2} = (\boldsymbol{A}^{1/2})^{-1}$. Let

$$\boldsymbol{P}_1 = \boldsymbol{C}(\boldsymbol{\psi},F_{\boldsymbol{\theta}})^{-1/2} \boldsymbol{Z}_n, \quad \boldsymbol{P}_2 = \boldsymbol{V}(\boldsymbol{\psi},F_{\boldsymbol{\theta}})^{-1/2}_{(22)} \boldsymbol{T}_{n(2)}.$$

By taking the inner product of the vectors $\boldsymbol{P}_1$ and $\boldsymbol{P}_2$, we obtain $\boldsymbol{P}_1^\top \boldsymbol{P}_2 = \boldsymbol{Z}_n^\top \boldsymbol{C}(\boldsymbol{\psi},F_{\boldsymbol{\theta}})^{-1/2} \boldsymbol{V}(\boldsymbol{\psi},F_{\boldsymbol{\theta}})^{-1/2}_{(22)} \boldsymbol{T}_{n(2)}$. Since

$$\boldsymbol{V}(\boldsymbol{\psi},F_{\boldsymbol{\theta}})^{-1/2}_{(22)} = \boldsymbol{C}(\boldsymbol{\psi},F_{\boldsymbol{\theta}})^{-1/2} \boldsymbol{M}(\boldsymbol{\psi},F_{\boldsymbol{\theta}})_{(22.1)},$$

it follows that

$$\boldsymbol{P}_1^\top \boldsymbol{P}_2 = \boldsymbol{Z}_n^\top \boldsymbol{C}(\boldsymbol{\psi},F_{\boldsymbol{\theta}})^{-1} \boldsymbol{M}(\boldsymbol{\psi},F_{\boldsymbol{\theta}})_{(22.1)} \boldsymbol{T}_{n(2)}.$$

On the basis of the above expression, we shall define the new robust test statistic, which is named as the *gradient-type test statistic*. Note that the above procedure is analogous to that of Terrell [6], who proposed the classical gradient statistic. We have the following definition.

Definition 5.1. The gradient-type statistic to test the composite null hypothesis $\mathcal{H}_0 : \boldsymbol{\theta}_{(2)} = \boldsymbol{\theta}_{0(2)}$ against $\mathcal{H}_\text{a} : \boldsymbol{\theta}_{(2)} \neq \boldsymbol{\theta}_{0(2)}$ in general parametric models is

$$G_n^2 = \boldsymbol{Z}_n^\top \boldsymbol{C}(\boldsymbol{\psi},F_{\boldsymbol{\theta}})^{-1} \boldsymbol{M}(\boldsymbol{\psi},F_{\boldsymbol{\theta}})_{(22.1)} \boldsymbol{T}_{n(2)}. \tag{5.3}$$

Remark. The $q \times q$ positive definite matrix

$$C(\boldsymbol{\psi}, F_{\boldsymbol{\theta}})^{-1} M(\boldsymbol{\psi}, F_{\boldsymbol{\theta}}) \quad (22.1)$$

in Eq. (5.3) can be estimated consistently by replacing $\boldsymbol{\theta}$ with $\boldsymbol{T}_n^{\omega}$, as in the case of the score-type statistic. □

The new robust test statistic in Eq. (5.3) is not complicated to be computed. The Wald-type statistic W_n^2 involves M-estimators only under the full model, whereas the score-type statistic R_n^2 considers M-estimators only under the reduced model. On the other hand, unlike its progenitors (the Wald- and score-type test statistics), the gradient-type statistic G_n^2 involves M-estimators under the full and reduced models, defined by Eqs. (5.1) and (5.2), respectively; that is, it contemplates the null and alternative hypotheses in its computation.

Next, we assume the conditions for the existence, consistency, and asymptotic normality of M-estimators as given by (A.1)–(A.9) in Heritier and Ronchetti [41]. These conditions have been studied by Clarke [44], among others. For the convenience of readers, we report these conditions in what follows. Let F be any arbitrary distribution on $\mathbb{R}$. Define $\boldsymbol{K}_F(\boldsymbol{\theta}) = \int \boldsymbol{\psi}(z, \boldsymbol{\theta})\, dF(z)$. We have

(A.1) $\boldsymbol{K}_F(\boldsymbol{\theta})$ exists at least on a (nondegenerate) open set $\mathcal{X} \subset \boldsymbol{\Theta}$.
(A.2) There exists $\boldsymbol{\theta}_*$ in $\mathcal{X}$ satisfying $\int \boldsymbol{\psi}(z, \boldsymbol{\theta}_*)\, dF(z) = \mathbf{0}_p$.
(A.3) $\int \boldsymbol{\psi}(z, \boldsymbol{\theta})\, dF_{\boldsymbol{\theta}}(z) = \mathbf{0}_p$ (Fisher consistency).
(A.4) $\boldsymbol{\psi}(z, \boldsymbol{\theta})$ is a $p \times 1$ vector function that is continuous and bounded on $\mathcal{Z} \times \mathcal{D}$, where $\mathcal{D}$ is some nondegenerate compact interval containing $\boldsymbol{\theta}_*$ (Fréchet differentiability).
(A.5) $\boldsymbol{\psi}(z, \boldsymbol{\theta})$ is locally Lipschitz in $\boldsymbol{\theta}$ about $\boldsymbol{\theta}_*$ in the sense that for some constant α,

$$\|\boldsymbol{\psi}(z, \boldsymbol{\theta}) - \boldsymbol{\psi}(z, \boldsymbol{\theta}_*)\| < \alpha \|\boldsymbol{\theta} - \boldsymbol{\theta}_*\|$$

uniformly in $z \in \mathcal{Z}$ and for all $\boldsymbol{\theta}$ in a neighborhood of $\boldsymbol{\theta}_*$.
(A.6) The generalized Jacobian $\partial \boldsymbol{K}_F(\boldsymbol{\theta})$ is of maximal rank at $\boldsymbol{\theta} = \boldsymbol{\theta}_*$.
(A.7) Given $\beta > 0$, there exists $\varepsilon > 0$ such that for all distributions in a ε neighborhood of F, $\sup_{\boldsymbol{\theta} \in \mathcal{D}} \|\boldsymbol{K}_G(\boldsymbol{\theta}) - \boldsymbol{K}_F(\boldsymbol{\theta})\| < \beta$ and $\partial \boldsymbol{K}_G(\boldsymbol{\theta}) \subset \partial \boldsymbol{K}_F(\boldsymbol{\theta}) + \beta \boldsymbol{B}$, uniformly in $\boldsymbol{\theta} \in \mathcal{D}$, where $\boldsymbol{B}$ is the unit ball of $p \times p$ matrices.
(A.8) $\boldsymbol{K}_F(\boldsymbol{\theta})$ has at least a continuous derivative $(\partial/\partial\boldsymbol{\theta})\boldsymbol{K}_F(\boldsymbol{\theta})$ at $\boldsymbol{\theta} = \boldsymbol{\theta}_*$.

(A.9) $(\partial\boldsymbol{\psi}/\partial\boldsymbol{\theta})(z,\boldsymbol{\theta})$ exists for $\boldsymbol{\theta} = \boldsymbol{\theta}_*$ and almost everywhere in a neighborhood of $\boldsymbol{\theta}_*$. Moreover, for all $\boldsymbol{\theta}$ in this neighborhood, there exists an integrable function $v(z)$ such that

$$\|(\partial\boldsymbol{\psi}/\partial\boldsymbol{\theta})(z,\boldsymbol{\theta})\| \leq v(z)$$

almost everywhere.

Conditions (A.1) and (A.2) ensure the existence of the functional $\boldsymbol{\theta}_* = \boldsymbol{T}(F)$ that defines an M-estimator through $\boldsymbol{T}(F^{(n)})$, where $F^{(n)}$ is the empirical distribution function. Condition (A.3) is a standard condition in the robustness literature and states the Fisher consistency of M-estimators. For a bounded $\boldsymbol{\psi}$-function, this implies consistency. Conditions (A.4)–(A.8) guarantee Fréchet differentiability for M-estimators. Fréchet differentiability is a strong form of differentiability that implies, for instance, the existence of the IF and the asymptotic normality for the corresponding estimator. These weak conditions include the case where the $\boldsymbol{\psi}$-function is not differentiable and are satisfied by the common choices of robust M-estimators. In particular, boundedness for $\boldsymbol{\psi}$ as required by (A.4) implies boundedness of the IF of the corresponding M-estimator. This is a key robustness condition. Finally, notice that if $\boldsymbol{K}_F(\boldsymbol{\theta})$ is continuously differentiable in $\boldsymbol{\theta}$, then (A.6) is equivalent to require that $\int(\partial\boldsymbol{\psi}/\partial\boldsymbol{\theta})(z,\boldsymbol{\theta}_*)\,dF(z)$ is nonsingular. The above comments were taken from Heritier and Ronchetti [41, p. 902].

The following theorem establishes the asymptotic distribution of the gradient-type test statistic given in Eq. (5.3).

Theorem 5.1. *Suppose that the conditions (A.1)–(A.9) are satisfied. Let* $\mathcal{H}_{\text{an}} : \boldsymbol{\theta}_{(2)} = n^{-1/2}\boldsymbol{\mu}$ *be a sequence of contiguous alternative hypotheses, where* $\boldsymbol{\mu}$ *is any vector in* $\mathbb{R}^q$ *such that* $(\boldsymbol{\theta}_{(1)}^\top, n^{-1/2}\boldsymbol{\mu}^\top)^\top$ *still belongs to* $\boldsymbol{\Theta}$, *and* $\boldsymbol{\theta}_{(1)} = \boldsymbol{\theta}_{0(1)}$. *Then, under the sequence of contiguous alternatives* $\mathcal{H}_{\text{an}}$, *the gradient-type statistic* nG_n^2 *has asymptotically a noncentral* $\chi^2_{q,\lambda}$ *distribution with noncentrality parameter* $\lambda = \boldsymbol{\mu}^\top \boldsymbol{V}(\boldsymbol{\psi}, F_{\boldsymbol{\theta}_0})^{-1}_{(22)}\boldsymbol{\mu}$, *where* $\boldsymbol{\theta}_0 = (\boldsymbol{\theta}_{0(1)}^\top, \boldsymbol{\theta}_{0(2)}^\top)^\top$.

Proof. We take into account the conditions (A.1)–(A.8). Let $F^{(n)}$ be the empirical distribution function. Conditions (A.1) and (A.2) ensure the existence of the functional $\boldsymbol{\theta}_* = \boldsymbol{T}(F)$ that defines an M-estimator through $\boldsymbol{T}_n = \boldsymbol{T}(F^{(n)})$. For any $\boldsymbol{\theta} \in \boldsymbol{\Theta}$, define $\boldsymbol{L}_n(\boldsymbol{\theta}) = n^{-1/2}\sum_{l=1}^{n}\boldsymbol{\psi}(z_l,\boldsymbol{\theta})$. For simplicity, let us denote the matrix $\boldsymbol{M}(\boldsymbol{\psi}, F_{\boldsymbol{\theta}})$ by $\boldsymbol{M}(\boldsymbol{\theta})$, the matrix

$M(\psi, F_\theta)_{(22.1)}$ by $M(\theta)_{(22.1)}$, and the matrix $C(\psi, F_\theta)$ by $C(\theta)$. By using a von Mises expansion of $T(F^{(n)})$, it follows that

$$\sqrt{n}(T_n - \theta) = M(\theta)^{-1} L_n(\theta) + O_p(n^{-1/2}),$$

which exists by (A.1)–(A.8). In particular,

$$\sqrt{n}(T_n - T_n^\omega) = M(T_n^\omega)^{-1} L_n(T_n^\omega) + O_p(n^{-1/2}).$$

Since $L_n(T_n^\omega)_{(1)} = n^{-1/2}\sum_{l=1}^n \psi(z_l, T_n^\omega)_{(1)} = \mathbf{0}_{p-q}$ by the definition of $T_n^\omega = (T_{n(1)}^{\omega\top}, \theta_{0(2)}^\top)^\top$ in Eq. (5.2), we have that

$$\sqrt{n}(T_{n(2)} - \theta_{0(2)}) = M(T_n^\omega)^{-1}_{(22.1)} L_n(T_n^\omega)_{(2)} + O_p(n^{-1/2}).$$

Now, using the fact that $Z_n = O_p(n^{-1/2})$ and by noting that $n^{-1/2}L_n(T_n^\omega)_{(2)} = Z_n$, the gradient-type statistic can be expressed as

$$\begin{aligned} G_n^2 &= Z_n^\top C(T_n^\omega)^{-1} M(T_n^\omega)_{(22.1)}[M(T_n^\omega)^{-1}_{(22.1)} Z_n + O_p(n^{-1})] \\ &= Z_n^\top C(T_n^\omega)^{-1} Z_n + O_p(n^{-3/2}). \end{aligned}$$

Since $C(T_n^\omega)$ is a consistent estimate of $C(\theta)$ at the model, we can write $nG_n^2 = nR_n^2 + O_p(n^{-1/2}) = nR_n^2 + o_p(1)$. It follows from Heritier and Ronchetti [41] that under $\mathcal{H}_{\text{an}}$, the statistic nR_n^2 has asymptotically a noncentral $\chi^2_{q,\lambda}$ distribution with noncentrality parameter $\lambda = \mu^\top V(\psi, F_{\theta_0})^{-1}_{(22)}\mu$. Therefore, nG_n^2 has the same asymptotic distribution as that of nR_n^2 under $\mathcal{H}_{\text{an}}$ and hence the result holds. □

An immediate consequence of Theorem 5.1 is the following theorem.

Theorem 5.2. *Assume conditions (A.1)–(A.9). Under the composite null hypothesis $\mathcal{H}_0 : \theta_{(2)} = \theta_{0(2)}$, the gradient-type statistic nG_n^2 has asymptotically a central χ^2_q distribution.*

Theorems 5.1 and 5.2 reveal that the Wald-, score-, and gradient-type test statistics have the same asymptotic distributional properties under either the null or contiguous alternatives hypotheses. Further, Theorems 5.1 and 5.2 also show that the robust gradient-type statistic has the same asymptotic distributional properties (under the null or alternative hypothesis) as its classical counterpart (see [12]).

The following theorem is related to the positiveness of the robust statistic defined in Eq. (5.3), since it is not transparently non-negative. We begin by observing that it is not possible to guarantee the positiveness without imposing many restrictions on the ψ-function. Nevertheless, we can prove a fairly general result that ensure that, for large sample size n, we have a high probability of having a strictly positive statistic G_n^2.

We initially prove a lemma. Given $\xi > 0$, let $\epsilon_n(\xi) = F_n^-(\xi/2)$, where F_n stands for the distribution function of nR_n^2, and F_n^- is the generalized inverse of F_n; that is, for a distribution function F, $F^-(x) = \sup\{t; F(t) \leq x\}$. Since $R_n^2 > 0$ with probability 1, it is clear that $\epsilon_n(\xi) > 0$.

Lemma 5.1. *Let $\xi > 0$ and $\epsilon(\xi) := \inf_{n\geq 1} \epsilon_n(\xi)$. Then, $\epsilon(\xi) > 0$.*

Proof. Suppose $\inf_{n\geq 1} \epsilon_n(\xi) = 0$. Then, we can find a subsequence $(n_k)_{k\geq 1}$ such that

$$\epsilon_{n_k} \overset{k\to\infty}{\longrightarrow} 0.$$

Thus, given any $x > 0$, we have that, for sufficiently large k, $\epsilon_{n_k} < x$. Therefore, we have that $\epsilon_{n_k} + (x - \epsilon_{n_k})/2 < x$, and from the definition of ϵ_{n_k}, $\Pr(nR_n^2 \leq \epsilon_{n_k} + (x - \epsilon_{n_k})/2) \geq \xi/2$. Then, if we denote by F_{n_k} the distribution function of $n_k R_{n_k}^2$, we obtain

$$\begin{aligned} F_{n_k}(x) &= \Pr(n_k R_{n_k}^2 \leq x) \\ &\geq \Pr(n_k R_{n_k}^2 \leq \epsilon_{n_k} + (x - \epsilon_{n_k})/2) \\ &\geq \frac{\xi}{2}. \end{aligned}$$

We then conclude that, $\forall x > 0$,

$$\limsup_{k\to\infty} F_{n_k}(x) \geq \frac{\xi}{2}.$$

Since nR_n^2 converges in distribution to a χ^2 distribution (central or non-central depending on whether we are assuming the null or alternative hypothesis), if we denote by F its distribution function, we have that, from Portmanteau's theorem, for every $x > 0$,

$$F(x) \geq \frac{\xi}{2}.$$

This is obviously a contradiction, since F is continuous, and $F(0) = 0$. This contradiction shows that $\inf_{n\geq 1} \epsilon_n(\xi) > 0$. □

Theorem 5.3. *Assume conditions (A.1)–(A.9). Let* $A_n = \{\omega \in \Omega; G_n^2(\omega) > 0\}$. *Then, under the composite null hypothesis* $\mathcal{H}_0 : \boldsymbol{\theta}_{(2)} = \boldsymbol{\theta}_{0(2)}$, *or under the sequence of contiguous alternatives* $\mathcal{H}_{\text{an}}$ *given in Theorem 5.1, the gradient-type statistic* G_n^2 *satisfies*

$$\lim_{n\to\infty} \Pr(A_n) = 1.$$

Proof. Recall that $A_n = \{\omega \in \Omega; nG_n^2(\omega) > 0\}$. Suppose that we have

$$\lim_{n\to\infty} \Pr(A_n) \nrightarrow 1.$$

This means that we can find $\xi > 0$ and a subsequence $(n_k)_{k\geq 1}$ such that $\Pr(A_{n_k}) < 1 - \xi$, or equivalently $\Pr(A_{n_k}^c) \geq \xi$; that is, for every k,

$$\Pr(n_k G_{n_k}^2 \leq 0) \geq \xi.$$

Let $\epsilon_{n_k}(\xi) = F_n^-(\xi/2)$, where F_n is the distribution function of nR_n^2. Then, from Lemma 5.1, $\epsilon(\xi) := \inf_{k\geq 1} \epsilon_{n_k}(\xi) > 0$. Recall from the proof of Theorem 5.1 that

$$nG_n^2 = nR_n^2 + O_p(n^{-1/2}),$$

or equivalently that

$$n(G_n^2 - R_n^2) \xrightarrow{\mathbb{P}} 0.$$

Note also that if $nG_n^2 \leq 0$ and $|nG_n^2 - nR_n^2| < \epsilon$, then $nR_n^2 < \epsilon$. This implies that

$$\Pr(n_k R_{n_k}^2 < \epsilon(\xi)) \geq \Pr(|n_k R_{n_k}^2 - n_k G_{n_k}^2| < \epsilon(\xi), n_k G_{n_k}^2 \leq 0).$$

Note that if $\Pr(A_n) \to 1$, then for every measurable B_n, we have $\Pr(A_n \cup B_n) \to 1$, and since $\Pr(B_n) = \Pr(A_n \cup B_n) - \Pr(A_n) + \Pr(A_n \cap B_n)$, it follows that $\limsup_{n\to\infty} \Pr(B_n) = \limsup_{n\to\infty} \Pr(A_n \cap B_n)$. Thus,

$$\begin{aligned}
&\limsup_{k\to\infty} \Pr(n_k R_{n_k}^2 < \epsilon(\xi)) \\
&\quad \geq \limsup_{k\to\infty} \Pr(|n_k R_{n_k}^2 - n_k G_{n_k}^2| < \epsilon(\xi), n_k G_{n_k}^2 \leq 0) \\
&\quad = \limsup_{k\to\infty} \Pr(n_k G_{n_k}^2 \leq 0) \\
&\quad \geq \xi,
\end{aligned}$$

where we used the fact that $\Pr(|n_k R^2_{n_k} - n_k G^2_{n_k}| < \epsilon(\xi)) \to 1$, since $n(G^2_n - R^2_n) \xrightarrow{\mathbb{P}} 0$. On the other hand, from the very definition of $\epsilon(\xi)$, we have

$$\limsup_{k\to\infty} \Pr(n_k R^2_{n_k} < \epsilon(\xi)) \leq \frac{\xi}{2}.$$

This contradiction proves the theorem. □

Let $\psi(z,\boldsymbol{\theta})$ be the derivative of $\rho(z,\boldsymbol{\theta})$ about $\boldsymbol{\theta}$ if the derivative exists and 0 otherwise. We arrive at the following theorem.

Theorem 5.4. *Let $\rho(z,\boldsymbol{\theta})$ be a convex function and differentiable at some $\boldsymbol{\theta} \in \boldsymbol{\Theta}$ such that $(\partial\rho/\partial\boldsymbol{\theta})(z,\boldsymbol{\theta}) = \psi(z,\boldsymbol{\theta})$. Then $G^2_n \geq 0$.*

Proof. The proof is similar to that of Theorem 2 in Terrell [6, p. 208]. □

Remark. When $\psi(z,\boldsymbol{\theta})$ is the score function, the gradient-type test statistic in Eq. (5.3) reduces simply to $G^2_n = \boldsymbol{Z}_n^\top \boldsymbol{T}_{n(2)}$, which is the classical gradient test statistic, as expected. □

5.3 ROBUSTNESS PROPERTIES

Following Heritier and Ronchetti [41], the robustness properties of the gradient-type test can be investigated by showing that a small amount of contamination at a point z has bounded influence on the asymptotic level and power of the test, which ensures the local stability of the test. The global reliability (or robustness against large deviations) could be measured by the breakdown point as defined in He et al. [46]. However, according to Cantoni and Ronchetti [47], small deviations are probably the main concern at the inference stage of a statistical analysis. We follow these authors and focus here on small deviations to study the robustness properties of the gradient-type test.

From Heritier and Ronchetti [41], the Wald- and score-type statistics can be written as functionals of the empirical distribution $F^{(n)}$ in the forms $W^2_n = W^2(F^{(n)})$ and $R^2_n = R^2(F^{(n)})$, where $W^2(F) = \boldsymbol{T}(F)^\top_{(2)} \boldsymbol{V}(\psi, F_{\boldsymbol{\theta}})^{-1}_{(22)} \boldsymbol{T}(F)_{(2)}$, $R^2(F) = \boldsymbol{Z}(F)^\top \boldsymbol{C}(\psi, F_{\boldsymbol{\theta}})^{-1} \boldsymbol{Z}(F)$, and $\boldsymbol{T}(F)$ and $\boldsymbol{Z}(F)$ are the functionals associated to $\boldsymbol{T}_n$ and $\boldsymbol{Z}_n$, respectively. To study the robustness properties of these tests, Heritier and Ronchetti [41] expressed the functional defining these test statistics as quadratic forms; that is, $W^2(F) = \boldsymbol{U}_{\mathrm{W}}(F)^\top \boldsymbol{U}_{\mathrm{W}}(F)$ and $R^2(F) = \boldsymbol{U}_{\mathrm{R}}(F)^\top \boldsymbol{U}_{\mathrm{R}}(F)$, where $\boldsymbol{U}_{\mathrm{W}}(F) = \boldsymbol{V}(\boldsymbol{\psi}, F_{\boldsymbol{\theta}})_{(22)}^{-1/2} \boldsymbol{T}(F)_{(2)}$ and $\boldsymbol{U}_{\mathrm{R}}(F) = \boldsymbol{C}(\boldsymbol{\psi}, F_{\boldsymbol{\theta}})^{-1/2} \boldsymbol{Z}(F)$.

To study the robustness properties of the gradient-type test, we also express the functional defining the statistic in Eq. (5.3) as quadratic form. From the definition of the gradient-type statistic, we have that $G_n^2 = [\boldsymbol{C}(\boldsymbol{\psi}, F_{\boldsymbol{\theta}})^{-1/2} \boldsymbol{Z}_n]^\top \boldsymbol{C}(\boldsymbol{\psi}, F_{\boldsymbol{\theta}})^{-1/2} \boldsymbol{M}(\boldsymbol{\psi}, F_{\boldsymbol{\theta}})_{(22.1)} \boldsymbol{T}_{n(2)}$. Since $\boldsymbol{C}(\boldsymbol{\psi}, F_{\boldsymbol{\theta}})^{-1/2} \boldsymbol{M}(\boldsymbol{\psi}, F_{\boldsymbol{\theta}})_{(22.1)} = \boldsymbol{V}(\boldsymbol{\psi}, F_{\boldsymbol{\theta}})_{(22)}^{-1}$, it follows that $G_n^2 = \boldsymbol{U}_{\mathrm{R}}(F^{(n)})^\top \boldsymbol{U}_{\mathrm{W}}(F^{(n)})$ and hence $G^2(F) = \boldsymbol{U}_{\mathrm{R}}(F)^\top \boldsymbol{U}_{\mathrm{W}}(F)$; that is,

$$G_n^2 = G^2(F^{(n)}) = \boldsymbol{U}_{\mathrm{R}}(F^{(n)})^\top \boldsymbol{U}_{\mathrm{W}}(F^{(n)}).$$

From the proof of Theorem 5.1, we have that

$$nG_n^2 = n\boldsymbol{U}_{\mathrm{R}}(F^{(n)})^\top \boldsymbol{U}_{\mathrm{R}}(F^{(n)}) + o_p(1),$$

and hence the gradient-type statistic is asymptotically equivalent to the positive definite quadratic form defined by the functional $\boldsymbol{U}_{\mathrm{R}}(F)$. Note that the gradient-type test statistic can be written (at least asymptotically) as a simple quadratic form of the functional $\boldsymbol{U}_{\mathrm{R}}(F)$. Then we can make use of the results for the score-type test derived by Heritier and Ronchetti [41] to analyze the asymptotic local stability properties of the gradient-type test.

Consider the ε-contamination $F_{\varepsilon,n} = (1 - \varepsilon/\sqrt{n})F_{\boldsymbol{\theta}_0} + (\varepsilon/\sqrt{n})K$, where K is an arbitrary distribution. Then, if we denote by $\alpha(F_{\varepsilon,n})$ the level of the test under the ε-contamination $F_{\varepsilon,n}$ and by α_0 the nominal level $\alpha(F_{\boldsymbol{\theta}_0})$, it follows from Heritier and Ronchetti [41] that

$$\lim_{n\to\infty} \alpha(F_{\varepsilon,n}) = \alpha_0 + \varepsilon^2 \kappa \left\| \int \mathrm{IF}(z; \boldsymbol{U}_{\mathrm{R}}, F_{\boldsymbol{\theta}_0})\, dK(z) \right\|^2 + o(\varepsilon^2),$$

where $\kappa = -(\partial/\partial\lambda) H_q(\eta_{1-\alpha_0}; \lambda)|_{\lambda=0} = (1-\alpha_0)/2 - H_{q+2}(\eta_{1-\alpha_0}; 0)/2$, $H_q(\eta_{1-\alpha_0}; \lambda)$ is the cumulative distribution function of a noncentral $\chi^2_{q,\lambda}$ distribution with q degrees of freedom and noncentrality parameter λ, $\eta_{1-\alpha_0}$ is the $1-\alpha_0$ quantile of the central χ^2_q distribution, and $\boldsymbol{U}_{\mathrm{R}}$ is the functional defining the quadratic form of the score-type test statistic. A similar result can be obtained for the power, showing that the asymptotic power is stable under contamination.

The above result derived by Heritier and Ronchetti [41] shows that the proper quantity to bound to have a stable level in a neighborhood around the hypothesis is the IF of the functional $\boldsymbol{U}_{\mathrm{R}}$. According to Heritier and Ronchetti [41], it is necessary to choose a contamination which converges to zero at the same rate as the sequence of contiguous alternatives converges to the null hypothesis in order to avoid overlapping between the neighborhood of the hypothesis and that of the alternative. In particular, we choose a point mass contamination $K(z) = \Lambda_z$:

$$F^L_{\varepsilon,n} = \left(1 - \frac{\varepsilon}{\sqrt{n}}\right) F_{\boldsymbol{\theta}_0} + \frac{\varepsilon}{\sqrt{n}} \Lambda_z.$$

We have the following theorem.

Theorem 5.5. *Under conditions (A.1)–(A.9), and for any M-estimator* $\boldsymbol{T} = (\boldsymbol{T}_{(1)}^\top, \boldsymbol{T}_{(2)}^\top)^\top$ *with bounded influence function of the second component* $\boldsymbol{T}_{(2)}$, *the asymptotic level of the robust gradient-type test under a point mass contamination is given by*

$$\lim_{n\to\infty} \alpha(F^L_{\varepsilon,n}) = \alpha_0 + \varepsilon^2 \kappa \, \mathrm{IF}(z; \boldsymbol{T}_{(2)}, F_{\boldsymbol{\theta}_0})^\top \times \boldsymbol{V}_{(22)}(\boldsymbol{T}_{(2)}, F_{\boldsymbol{\theta}_0})^{-1} \mathrm{IF}(z; \boldsymbol{T}_{(2)}, F_{\boldsymbol{\theta}_0}) + o(\varepsilon^2). \tag{5.4}$$

Proof. The proof follows immediately from the results for the score-type test in Heritier and Ronchetti [41] by noting that $G_n^2 = \boldsymbol{U}_{\mathrm{R}}(F^{(n)})^\top \boldsymbol{U}_{\mathrm{R}}(F^{(n)})$ up to an error of order $o_p(n^{-1})$. □

As noted by Heritier and Ronchetti [41],

$$[\mathrm{IF}(z; \boldsymbol{T}_{(2)}, F_{\boldsymbol{\theta}_0})^\top \boldsymbol{V}_{(22)}(\boldsymbol{T}_{(2)}, F_{\boldsymbol{\theta}_0})^{-1} \mathrm{IF}(z; \boldsymbol{T}_{(2)}, F_{\boldsymbol{\theta}_0})]^{1/2},$$

is the self-standardized IF of the M-estimator $\boldsymbol{T}_{(2)}$ and hence it has to be bound to obtain a robust gradient-type test. In other words, a bounded-influence M-estimator $\boldsymbol{T}_{(2)}$ in Eq. (5.4) ensures a bound on the asymptotic level of the robust gradient-type test under contamination. Then an optimally bounded-influence gradient-type test can be obtained by using an optimally robust self-standardized estimator $\boldsymbol{T}_{(2)}$ in Eq. (5.4). However, we consider a simpler gradient-type test that bounds the self-standardized IF given in Eq. (5.4) following Heritier and Ronchetti [41], who proposed the robust bounded-influence Wald- and score-type tests. Let c ($c > \sqrt{q}$) be a given bound on the self-standardized IF in Eq. (5.4). Consider

$$\boldsymbol{\psi}_c^*(z, \boldsymbol{\theta}) = w_c^*(z, \boldsymbol{\theta})[\boldsymbol{s}(z, \boldsymbol{\theta}) - \boldsymbol{a}(\boldsymbol{\theta})], \tag{5.5}$$

where $s(z,\theta)$ is the score vector,

$$w_c^*(z,\theta) = \min\left\{1, \frac{c}{\|b(\theta)_{(2)}\|}\right\},$$

and $b(\theta) = (b(\theta)_{(1)}^\top, b(\theta)_{(2)}^\top)^\top = A(\theta)[s(z,\theta) - a(\theta)]$. So, the gradient-type test will be based on the ψ-function in Eq. (5.5). An M-estimator obtained by using the ψ-function in Eq. (5.5) is known as the standardized optimal bias-robust estimator (OBRE). This estimator is optimal in the sense that it is the M-estimator which minimizes the trace of the asymptotic covariance matrix under constraint that it has a bounded IF (see [48]). Here, the p-vector $a(\theta)$ is determined implicitly by the equation

$$a(\theta) = \frac{\int s(z,\theta) w_c^*(z,\theta)\, dF_\theta(z)}{\int w_c^*(z,\theta)\, dF_\theta(z)}, \tag{5.6}$$

whereas the lower triangular $p \times p$ matrix $A(\theta)$ is determined implicitly from

$$A(\theta)^\top A(\theta) = \left[\int [s(z,\theta) - a(\theta)][s(z,\theta) - a(\theta)]^\top \times w_c^*(z,\theta)^2\, dF_\theta(z)\right]^{-1}. \tag{5.7}$$

Eqs. (5.6) and (5.7) are given in Hampel et al. [48, p. 254]. According to Heritier and Ronchetti [41], the quantity $\|(A(\theta)\psi_c^*(z,\theta))_{(2)}\|$ is the self-standardized IF for the estimator $T_{(2)}$ defined by the ψ-function in Eq. (5.5), where $(d)_{(2)}$ stands for the last $q \leq p$ components of the p-vector d, and hence the weight $w_c^*(z,\theta)$ has the effect of bounding this quantity by c. The same weight is used on the IF of the nuisance component, and this simplifies the form of the ψ-function. Therefore, the robust bounded-influence gradient-type test is defined from Eq. (5.3), considering the ψ-function given by Eqs. (5.5)–(5.7).

A brief comment on the tuning constant c is in order, since the ψ-function in Eq. (5.5) depends on c. For robust testing purposes, Ronchetti and Trojani [49] suggested that one can choose the constant which controls the maximal bias on the asymptotic level of the test in a neighborhood of the model. To this regard, one can take into account Theorem 5.5. Following Cantoni and Ronchetti [47] and take into account Theorem 5.5, the maximal level α of the robust gradient-type test in a neighborhood of the model of radius ε is given by

$$\alpha = \alpha_0 + \varepsilon^2 \kappa\, \gamma(T_{(2)}, F_{\theta_0})^2,$$

where $\gamma(\boldsymbol{T}_{(2)}, F_{\boldsymbol{\theta}_0}) = \sup_z \|\mathrm{IF}(z; \boldsymbol{T}_{(2)}, F_{\boldsymbol{\theta}_0})\|$. So, we can write $b = \varepsilon^{-1}\kappa^{-1/2}(\alpha - \alpha_0)^{1/2}$, were b is the bound on the IF of the estimator $\boldsymbol{T}_{(2)}$. Then, according to Cantoni and Ronchetti [47], for a fixed amount of contamination ε and by imposing a maximal error on the level of the test $\alpha - \alpha_0$, one can determine the bound b on the IF of the estimator, and hence the tuning constant by solving

$$b = \gamma(\boldsymbol{T}_{(2)}, F_{\boldsymbol{\theta}_0}) = \gamma_c$$

with respect to c.

Cantoni and Ronchetti [47] pointed out that one would have to choose a different value of c for each considered test and hence it is unreasonable from a practical point of view. On the other hand, they suggest to choose a global value of c by solving

$$b = \sup_z \|\mathrm{IF}(z; \boldsymbol{T}_{(2)}, F_{\boldsymbol{\theta}_0})\|,$$

based on the fact that $\gamma(\boldsymbol{T}_{(2)}, F_{\boldsymbol{\theta}_0}) = \sup_z \|\mathrm{IF}(z; \boldsymbol{T}_{(2)}, F_{\boldsymbol{\theta}_0})\| \leq \sup_z \|\mathrm{IF}(z; \boldsymbol{T}, F_{\boldsymbol{\theta}_0})\|$. Note that the solution depends on the unknown parameter $\boldsymbol{\theta}_0$, but Cantoni and Ronchetti [47] wrote "our experience shows that it does not vary much for different values of $\boldsymbol{\theta}$, so that one can safely plug-in a reasonable (robust) estimate."

5.4 ALGORITHM TO COMPUTE THE GRADIENT-TYPE STATISTIC

As noted earlier, the gradient-type statistic involves M-estimators under the full and reduced models, defined by Eqs. (5.1) and (5.2), respectively. Next, we summarize the algorithm for computing the robust bounded-influence gradient-type test statistic. The algorithm is as follows.

Step 1. Fix a precision threshold $\delta > 0$, and an initial starting point for the parameter vector $\boldsymbol{\theta}$. Fix the tuning constant c. Set $\boldsymbol{a} = \boldsymbol{0}_p$ and $\boldsymbol{A} = [\boldsymbol{J}(\boldsymbol{\theta})^{1/2}]^{-\top}$, where $\boldsymbol{J}(\boldsymbol{\theta}) = \int \boldsymbol{s}(z, \boldsymbol{\theta})\boldsymbol{s}(z, \boldsymbol{\theta})^\top \, dF_{\boldsymbol{\theta}}(z)$ is the Fisher information matrix for $\boldsymbol{\theta}$. The matrix $\boldsymbol{A}$ is chosen to be lower triangular.

Step 2. Solve the following two equations for $\boldsymbol{a}$ and $\boldsymbol{A}$:

$$\boldsymbol{a} = \frac{\int \boldsymbol{s}(z, \boldsymbol{\theta}) w_c^*(z, \boldsymbol{\theta}) \, dF_{\boldsymbol{\theta}}(z)}{\int w_c^*(z, \boldsymbol{\theta}) \, dF_{\boldsymbol{\theta}}(z)}, \quad \boldsymbol{A}^\top \boldsymbol{A} = \boldsymbol{M}_2^{-1};$$

where

$$M_k = \int [s(z,\theta) - a][s(z,\theta) - a]^\top w_c^*(z,\theta)^k \, dF_\theta(z), \quad k = 1,2.$$

The current values of θ, a, and A are used as starting values to solve the equations in this step.

Step 3. Compute M_1 using a and A from Step 2, and

$$\Delta\theta = M_1^{-1} \left[\frac{1}{n} \sum_{l=1}^{n} w_c^*(z_l,\theta)[s(z_l,\theta) - a] \right].$$

Step 4. If $\|\Delta\theta\| > \delta$, then $\theta \leftarrow \theta + \Delta\theta$ and return to Step 2; otherwise, stop.

The above algorithm computes the M-estimate under the full model, that is, T_n. The computation of the M-estimate under the reduced model follows the same algorithm, but with $\theta_{(2)} = \mathbf{0}_q$ and M_1^{-1} replaced by

$$\begin{bmatrix} M_{1(11)}^{-1} & \mathbf{0}_{p-q,q} \\ \mathbf{0}_{q,p-q} & \mathbf{0}_{q,q} \end{bmatrix}$$

in Step 3; and hence T_n^{ω} is computed. After computing the M-estimates T_n (full model) and T_n^{ω} (reduced model), one can compute the gradient-type test statistic given by Eq. (5.3).

It should be noted that the numerical integration can be avoided in Step 1 and in computing M_k (Steps 2 and 3) by replacing F_θ with its empirical cumulative distribution function. However, numerical integration must be used to compute a, otherwise the ψ-function in Eq. (5.5) will be satisfied by all estimates.

5.5 CLOSING REMARKS

From Eq. (5.3), it can be noted that the robust gradient-type test statistic is not complicated to be computed and it is a valuable complement to the classic gradient statistic. Also, it is more reliable in the presence of outlying points and other deviations from the assumed model than its classical counterpart. According to Dell'Aquila and Ronchetti [50], the general idea of robust statistics is to provide estimators and tests that are stable when the distributional assumptions are slightly different from the model assumptions. In the case of tests, this means, that the test should maintain approximately the correct size and not lose too much power when the empirical distribution differs (slightly) from the assumed model distribution. Still according to Dell'Aquila and Ronchetti [50], an empirical distribution may be viewed as a distribution in an "neighborhood" of the true distribution and one can therefore expect that in small samples a robust test has a stable size and power for a wide variety of distributions, while the performance of the classical tests will depend more heavily on the underlying distribution. Therefore, on the basis of the above comments, one should evaluate the size and power properties of the robust gradient-type test under various types of deviations from model assumptions for several class of parametric models by using Monte Carlo simulation experiments. From these numerical studies, one can verify the accuracy (in finite samples) of the gradient-type test that the uses the statistic (5.3) for testing hypothesis. This is worth investigating and could be considered in future research.

BIBLIOGRAPHY

[1] S. Wilks, The large-sample distribution of the likelihood ratio for testing composite hypothesis, Ann. Math. Stat. 9 (1938) 60–62.

[2] A. Wald, Tests of statistical hypothesis concerning several parameters when the number of observations is large, Trans. Am. Math. Soc. 54 (1943) 426–482.

[3] C. Rao, Large sample tests of statistical hypotheses concerning several parameters with applications to problems of estimation, Proc. Camb. Philos. Soc. 44 (1948) 50–57.

[4] C. Rao, Score test: historical review and recent developments, in: N. Balakrishnan, N. Kannan, H.N. Nagaraja (Eds.), Advances in Ranking and Selection, Multiple Comparisons, and Reliability, Birkhäuser, Boston, 2005.

[5] P. Sen, J. Singer, Large Sample Methods in Statistics: An Introduction with Applications, Chapman and Hall, New York, USA, 1993.

[6] G. Terrell, The gradient statistic, Comput. Sci. Stat. 34 (2002) 206–215.

[7] V. Muggeo, G. Lovison, The "three *plus* one" likelihood-based test statistics: unified geometrical and graphical interpretations, Am. Stat. 68 (2014) 302–306.

[8] A. Lemonte, S. Ferrari, Testing hypotheses in the Birnbaum-Saunders distribution under type-II censored samples, Comput. Stat. Data Anal. 55 (2011) 2388–2399.

[9] Z. Birnbaum, S. Saunders, A new family of life distributions, J. Appl. Probab. 6 (1969) 319–327.

[10] H. Ng, D. Kundu, N. Balakrishnan, Point and interval estimation for the two-parameter Birnbaum-Saunders distribution based on Type-II censored samples, Comput. Stat. Data Anal. 50 (2006) 3222–3242.

[11] W. Nelson, W. Meeker, Theory for optimum accelerated censored life tests for Weibull and extreme value distributions, Technometrics 20 (1978) 171–177.

[12] A. Lemonte, S. Ferrari, The local power of the gradient test, Ann. Inst. Stat. Math. 64 (2012) 373–381.

[13] P. Harris, H. Peers, The local power of the efficient score test statistic, Biometrika 67 (1980) 525–529.

[14] T. Hayakawa, The likelihood ratio criterion for a composite hypothesis under a local alternative, Biometrika 62 (1975) 451–460.

[15] D. Cox, N. Reid, Parameter orthogonality and approximate conditional inference, J. R. Stat. Soc. B 40 (1987) 1–39.

[16] A. Lemonte, Local power properties of some asymptotic tests in symmetric linear regression models, J. Stat. Plan. Inference 142 (2012) 1178–1188.

[17] P. McCullagh, J. Nelder, Generalized Linear Models, 2nd ed., Chapman and Hall, London, UK, 1989.

[18] A. Lemonte, Nonnull asymptotic distributions of the LR, Wald, score and gradient statistics in generalized linear models with dispersion covariates, Statistics 47 (2013) 1249–1263.

[19] T. Vargas, S. Ferrari, A. Lemonte, Gradient statistic: higher-order asymptotics and Bartlett-type correction, Electron. J. Stat. 7 (2013) 43–61.

[20] J. Ghosh, R. Mukerjee, Characterization of priors under which Bayesian and frequentist Bartlett corrections are equivalent in the multiparameter case, J. Multivar. Anal. 38 (1991) 385–393.

[21] T. Hayakawa, The likelihood ratio criterion and the asymptotic expansion of its distribution, Ann. Inst. Stat. Math. 29 (1977) 359–378.

[22] G. Cordeiro, S. Ferrari, A modified score test statistic having chi-squared distribution to order n^{-1}, Biometrika 78 (1991) 573–582.

[23] G. Hill, A. Davis, Generalized asymptotic expansions of Cornish-Fisher type, Ann. Math. Stat. 39 (1968) 1264–1273.

[24] J. Rieck, J. Nedelman, A log-linear model for the Birnbaum-Saunders distribution, Technometrics 33 (1991) 51–60.

[25] T. Vargas, S. Ferrari, A. Lemonte, Improved likelihood inference in generalized linear models, Comput. Stat. Data Anal. 74 (2014) 110–124.

[26] R. Freund, Regression with SAS with emphasis on PROC REG, in: Eighth Annual SAS Users Group International Conference, New Orleans, Louisiana, January 16–19, 1983.

[27] A. Lemonte, Local power of some tests in exponential family nonlinear models, J. Stat. Plan. Inference 141 (2011) 1981–1989.

[28] A. Lemonte, Nonnull asymptotic distributions of some criteria in censored exponential regression, Commun. Stat. Theory Methods 43 (2014) 3314–3328.

[29] A. Lemonte, S. Ferrari, Size and power properties of some tests in the Birnbaum-Saunders regression model, Comput. Stat. Data Anal. 55 (2011) 1109–1117.

[30] A. Lemonte, S. Ferrari, A note on the local power of the LR, Wald, score and gradient tests, Electron. J. Stat. 6 (2012) 421–434.

[31] A. Lemonte, S. Ferrari, Local power and size properties of the LR, Wald, score and gradient tests in dispersion model, Stat. Methodol. 9 (2012) 537–554.

[32] F. Medeiros, A. da Silva-Júnior, D. Valença, S. Ferrari, Testing inference in accelerated failure time models, Int. J. Stat. Probab. 3 (2014) 121–131.

[33] S. Ferrari, E. Pinheiro, Small-sample likelihood inference in extreme-value regression models, J. Stat. Comput. Simul. 84 (2014) 582–595.

[34] A. Lemonte, On the gradient statistic under model misspecification, Stat. Probab. Lett. 83 (2013) 390–398.

[35] K. Viraswami, N. Reid, Higher-order asymptotics under model misspecification, Can. J. Stat. 24 (1996) 263–278.

[36] K. Viraswami, N. Reid, A note on the likelihood-ratio statistic under model misspecification, Can. J. Stat. 26 (1998) 161–168.

[37] J. Stafford, A robust adjustment of the profile likelihood, Ann. Stat. 24 (1996) 336–352.

[38] R. Royall, T. Tsou, Interpreting statistical evidence by using imperfect models: robust adjusted likelihood functions, J. R. Stat. Soc. B 65 (2003) 391–404.

[39] T. Kent, Robust properties of likelihood ratio tests, Biometrika 69 (1982) 19–27.

[40] D. Fraser, On information in statistics, Ann. Math. Stat. 36 (1965) 890–896.

[41] S. Heritier, E. Ronchetti, Robust bounded-influence tests in general parametric models, J. Am. Stat. Assoc. 89 (1994) 897–904.

[42] C. Varin, N. Reid, D. Firth, An overview of composite likelihood methods, Stat. Sin. 21 (2011) 5–42.

[43] C. Kuş, A new lifetime distribution, Comput. Stat. Data Anal. 51 (2007) 4497–4509.

[44] B. Clarke, Nonsmooth analysis and Fréchet differentiability of M-functionals, Probab. Theory Relat. Fields 73 (1986) 197–209.

[45] P. Huber, E. Ronchetti, Robust Statistics, 2nd ed., Wiley, New York, USA, 2009.

[46] X. He, D. Simpson, S. Portnoy, Breakdown robustness of tests, J. Am. Stat. Assoc. 85 (1990) 446–452.

[47] E. Cantoni, E. Ronchetti, Robust inference for generalized linear models, J. Am. Stat. Assoc. 96 (2001) 1022–1030.

[48] F. Hampel, E. Ronchetti, P. Rousseeuw, W. Stahel, Robust Statistics: The Approach Based on Influence Functions, Wiley, New York, USA, 1986.

[49] E. Ronchetti, F. Trojani, Robust inference with GMM estimators, J. Econ. 101 (2001) 36–79.

[50] R. Dell'Aquila, E. Ronchetti, Robust tests of predictive accuracy, METRON LXII (2004) 161–184.

Printed in the United States
By Bookmasters